Mate Choice in Plants

MONOGRAPHS IN POPULATION BIOLOGY

EDITED BY ROBERT M. MAY

1. The Theory of Island Biogeography, by Robert H. MacArthur and Edward O. Wilson
2. Evolution in Changing Environments: Some Theoretical Explorations, by Richard Levins
3. Adaptive Geometry of Trees, by Henry S. Horn
4. Theoretical Aspects of Population Genetics, by Motoo Kimura and Tomoko Ohta
5. Populations in a Seasonal Environment, by Stephen D. Fretwell
6. Stability and Complexity in Model Ecosystems, by Robert M. May
7. Competition and the Structure of Bird Communities, by Martin L. Cody
8. Sex and Evolution, by George C. Williams
9. Group Selection in Predator-Prey Communities, by Michael E. Gilpin
10. Geographic Variation, Speciation, and Clines, by John A. Endler
11. Food Webs and Niche Space, by Joel E. Cohen
12. Caste and Ecology in the Social Insects, by George F. Oster and Edward O. Wilson
13. The Dynamics of Arthropod Predator-Prey Systems, by Michael P. Hassell
14. Some Adaptations of Marsh-Nesting Blackbirds, by Gordon H. Orians
15. Evolutionary Biology of Parasites, by Peter W. Price
16. Cultural Transmission and Evolution: A Quantitative Approach, by L. L. Cavalli-Sforza and M. W. Feldman
17. Resource Competition and Community Structure, by David Tilman
18. The Theory of Sex Allocation, by Eric L. Charnov
19. Mate Choice in Plants: Tactics, Mechanisms, and Consequences, by Mary F. Willson and Nancy Burley

Mate Choice in Plants

TACTICS, MECHANISMS, AND CONSEQUENCES

MARY F. WILLSON AND

NANCY BURLEY

PRINCETON UNIVERSITY PRESS

PRINCETON, NEW JERSEY

1983

Published by Princeton University Press,
41 William Street, Princeton, New Jersey
In the United Kingdom: Princeton University Press,
Guildford, Surrey

Library of Congress Cataloging in Publication Data will be found on the last printed page of this book
ISBN 0-691-08333-9
ISBN 0-691-08334-7 (pbk.)
This book has been composed in Linotron Baskerville
Clothbound editions of Princeton University Press books are printed on acid-free paper, and binding materials are chosen for strength and durability. Paperbacks, although satisfactory for personal collections, are not usually suitable for library rebinding.

Printed in the United States of America by
Princeton University Press, Princeton, New Jersey

It is a favourite popular delusion that the scientific inquirer is under a sort of moral obligation to abstain from going beyond that generalization of observed facts which is absurdly called "Baconian" induction. But any one who is practically acquainted with scientific work is aware that those who refuse to go beyond fact, rarely get as far as fact; and any one who has studied the history of science knows that almost every great step therein has been made by the "anticipation of Nature," that is, by the invention of hypotheses, which, though verifiable, often had very little foundation to start with; and, not unfrequently, in spite of a long career of usefulness, turned out to be wholly erroneous in the long run.

T. H. Huxley, 1896,
Method and Results; Essays.

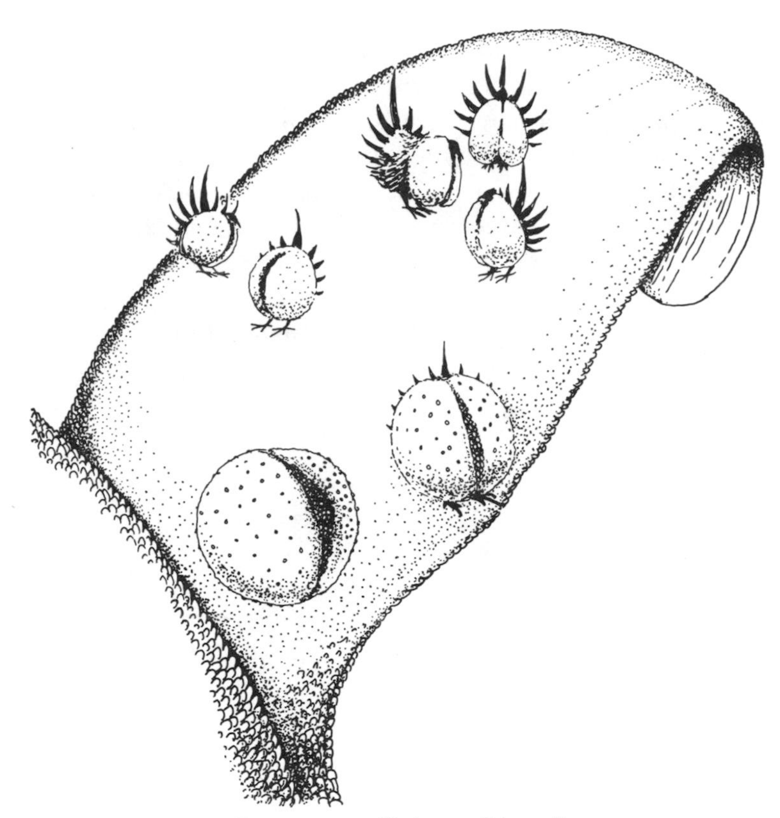

Frontispiece. "Lek on a Stigma."

Preface

This monograph attempts to address two audiences: botanists with interests in either population biology or suborganismal (developmental, physiological, and cellular) processes and that group of evolutionary biologists (almost all of whom are zoologists) whose interests center on intrapopulation interactions of a "sociobiological" nature. Because there is so little intellectual exchange between these groups, writing for both has proved challenging. We are aware that we are vulnerable to misinterpretation from both groups, particularly from botanists, because we apply a sociobiological perspective to materials notably bereft of this evolving body of theory. This book is about sociobotany.

As is common in sociobiological writing, our language sometimes appears anthropomorphic. Although plants are unable to make cognitive decisions, we are confident that they can and do make decisions routinely, using various physiological processes that are almost unexplored in this light. It makes little sense to avoid such simple descriptive terms as "choice," "decision," "conflict," and "tactic" just because humans engage in conflict and decision making in ways that seem distinct from most other species. Moreover, it would be awkwardly pretentious to coin new jargon that means the same thing but might prove more palatable simply because it is unfamiliar. What, for example, would be served by inventing a new word for "cuckoldry"? In many cases use of the semantic shorthand of sociobiology is a convenient way to avoid a long-winded circumlocution. Consider the term "tactic." The dictionary at hand defines a tactic as "an adroit device for accomplishing an end." In this monograph "tactic" refers to a means of increasing its owner's fitness, especially in response to a particular challenge from the biological environment. Tactics thus are adaptations, fashioned by natural selection, favoring the genetic bearers of superior adroit devices. Use of such a word in this way does not imply any conscious intent or planning; it is a *post facto* shorthand description of a response. There is no need to ascribe human versions of such words or attendant human emotions to any other organism; indeed, such ascription might be seen as a far greater anthropocentrism than the use of the words in the first place. Those uncomfortable with our use of language might wish to consult George Williams' discussion of "teleonomy" in his classic book, *Adaptation and Natural Selection* (1966).

Higher plants can be distinguished from most animals in several ways (for example, the immobility of adult individuals) that may have implications for understanding their sexual and social interactions. We draw analogies from animal behavior not to paper over the evident differences between plants and animals but to generate some new possible interpretations of certain fundamental processes.

Here we present a conceptual perspective on female choice of mates and to a lesser extent on inter-male competition; the perspective generates numerous hypotheses, many of which are testable. The purpose of this exercise is to permit the collection of data in new and meaningful ways. A novel perspective on what has been, historically, an independent area of investigation, such as plant embryology, leads to the formation of alternative explanatory hypotheses, an essential step in the proper pursuit of science (e.g., Platt, 1964). Our framework is summarized in a theory map (Figure 12) in which the relative importance, relationships, and centrality of the hypotheses are indicated.

Our discussion is aimed at exploring possible adaptive bases of some aspects of plant reproduction. We are deliberately speculative at many points in our discussion. This speculation allows us to elaborate upon the most basic and general hypotheses, to carry some of our ideas to a logical conclusion. Because of the limited amount of relevant information available, we cannot estimate what proportion of our hypotheses eventually might find support. If some of them lead to the conceptualization of superior hypotheses, their function will have been served.

Two potentially confusing terms used throughout this monograph may be profitably defined at the outset. "Hermaphrodite" is a term in the botanical literature used commonly to mean flowers with both male and female parts. We use it in its zoological sense, meaning an *individual* with both kinds of sex organs, whether or not they are in the same flower. This use of the word, by botanists, is becoming increasingly common. Further, we mean that both male and female parts are not only present but produce gametes, that the individuals can *function* as either or both sex(es). "Gymnosperm" was the collective name formerly applied to the kinds of plants now placed in the separate divisions Cycadophyta, Coniferophyta, Gnetophyta, Ginkgophyta (Bold *et al.*, 1980). We continue to use the collective term as a convenient means of comparing these plants with the angiosperms.

C. K. Augspurger shared in the fun of developing these ideas. R. Symanski contributed to the entire manuscript, especially to the conception and development of Chapter 5. Z. B. Carothers, J.M.J. de Wet, F. Meins, D. L. Nanney, G. S. Whitt, and J. H. Willis supplied entrees into a previously alien literature. W. G. Eberhard, D. A. Goldman, H. S. Horn, S. J. Kenny, R. A. Metcalf, D. Queller, D. W. Schemske, and G. C. Williams commented helpfully on the manuscript; R. M. May furnished a salubrious intellectual environment for refinement of the manuscript. Our biology librarians assisted graciously. P. W. Hedrick and L. Lionni contributed to development of vocabulary. R. A. von Neumann provided cheerful consultation.

Contents

Mate Choice in Plants

CHAPTER ONE

Introduction and Theoretical Background

Animals employ many well-known means of choosing mates. Many of these mechanisms are actively behavioral, an option that, in the usual sense, is not available to plants. Nevertheless, plants are not the totally passive and undiscriminating recipients of sperm that they are often thought to be. Although they can neither dance nor sing, plants possess a number of potential means of discriminating among pollen donors, and the donors possess several possible mechanisms for rendering themselves discriminable by pollen receivers. As far as we know, the events and processes discussed in this book have, with few exceptions, not been considered in the light of mate choice, although the existence of botanical mate choice has been emphasized previously (Bateman, 1948; Janzen, 1977; Charnov, 1979; Willson, 1979).

We begin by presenting the theoretical backdrop. There are two major conceptual networks relevant to the ideas developed herein: those related to parent-offspring interactions (kin selection, parent-offspring conflict, and parental manipulation) and those dealing with sexual selection, particularly mate choice.

1.1 SEXUAL SELECTION

Darwin long ago (1871) described two aspects of sexual selection: (1) sexual competition among the members of one sex, typically male, for the chance to mate with individuals of the other sex, typically female, and (2) "epigamic" preferences of one sex, typically females, for particular mates. In 1948 Bateman made the issue more general, provided evidence from *Drosophila melanogaster*, and noted that sexual selection might also occur in plants. He found that for fruit flies the variance of male reproductive success (RS) far exceeded that of female RS, and that a male's RS was limited, not by his willingness or interest, but by the relative numbers of females he successfully courted (Payne, 1979; Wade, 1979; Wade and Arnold, 1980). He also found that a female's RS was limited, not by the number of males with which she copulated, but by the resouces she could mobilize for egg production.

The extent to which each sex experiences intrasexual competition and epigamic selection depends upon the mating system. One factor strongly related to the mating system is the pattern of parental investment (PI), which Trivers (1972) defined as any expenditure or risk taken by the parent for the benefit of an offspring that decreases the likelihood of success of future offspring. When females have high PI compared to males, they should be selective of their mates in order to avoid poor investments. In this case females are a limited resource for males and male-male competition for mates occurs. The intensity of sexual selection can be indexed by the disparity (as a ratio) of net RS of breeding males and females, where net RS is measured as benefit to the parent (in present offspring) minus the cost (in reduced numbers and/or quality of future offspring).

One aspect of sexual selection that often has both epigamic and intrasexual components is mating investment (MI: see Low, 1978; Alexander and Borgia, 1979), which is expenditure or risk taken to secure a partner that decreases the likelihood of future matings. Conceptually it is relatively easy to distinguish MI from PI; empirically the distinction is often much less clear. For example, males of many avian species defend territories both to acquire mates (MI) and to contribute to the growth and survival of offspring (PI). How could one quantify the relative proportions of MI and PI in such cases?

Sexual selection can exacerbate initial differences between male and female patterns of investment. Therefore some of the eventual differences between the sexes may seem to be the result and not the cause of sexual selection. As the degree of differential investment changes, however, the degree of intensity of sexual selection is affected in subsequent generations, and what was the effect of past selection is part of the cause of ensuing selection.

We interpret a number of pre- and postzygotic processes and phenomena in terms of sexual selection and, particularly, mate choice. Our use of the term "mate choice" to include events that occur after fertilization may generate disagreement. Some botanists and zoologists apparently feel that mate choice must include only events that occur prezygotically, or even more restrictedly (for animals), those that occur prior to sexual union. Two factors probably contribute to this position: 1) ambiguity in the meaning of the term "mate" and 2) lack of full consideration of the consequences of postzygotic activities on both intra- and intersexual components of sexual selection.

The term "mate" has multiple meanings in the biological literature. As a verb it can mean: 1) to form a bond for breeding or 2) to engage in the specific physical event that under certain conditions results in fertilization of eggs. (Copulation, for example, a frequent synonym for mating in the sense of sexual union, does not necessarily result in fertilization.) As a noun "mate" may refer to a bonded individual that has formed an alliance for breeding or to an individual with which one engages in sexual "union." In plants and in many animals no bonding occurs, however, and mating refers only to the events immediately surrounding sexual union. We offer a third definition: a mate is a partner accepted for shared parenthood of viable offspring. Our reasons for coining this definition may be best appreciated by example.

Among red-winged blackbirds, adult males compete for territories that are used by females for nesting and often for feeding during the breeding season (Orians, 1961, 1980). Females exercise mate choice, apparently selecting largely on the basis of the quality of the male territory (Wittenberger, 1976; Searcy, 1979a,b) but possibly also on male phenotypic characteristics (Weatherhead and Robertson, 1977, 1979; Yasukawa, 1981). A male may have one or more females resident on his territory. Prior to laying her eggs, a female copulates with the male on whose territory she resides. In addition, females at least sometimes copulate with other males (probably residents of nearby territories) as well (Bray *et al.*, 1975; Searcy, 1979b). We do not know the circumstances that affect the frequency of this female behavior, nor do we know how often "other males" fertilize one or more of the eggs in a clutch. For illustrative purposes, let us assume that such males are sometimes successful in fathering the "offspring" of cuckolded territory residents. Let us also assume that some phenomenon related to the timing of ovulation influences the probability of conception. A female then could "strategically" (but not with conscious intent) time her copulations to affect the probability of conception by the sperm of two or more males. Conceivably females might even pair with one male to obtain the best available nesting site but then look elsewhere for the best genes for their offspring. Selection pressure acting on males will tend to limit the success of this tactic but may not preclude it altogether.

The question of which male(s) constitute(s) a female's mate(s) under this circumstance is open to debate. By definition 1 above the male on whose territory she resides is her mate. By definition 2 all males with which she copulates qualify as mates. But only by definition 3 is the most important theoretical distinction made clear: which male(s) fathered the offspring? By our definition mate choice can occur prior to sexual union, after sexual union (for example, by the differential destruction of gametes), after fertilization (by the differential destruction of embryos, or abortion), and theoretically, at least in animals, up until the offspring have received their full complement of parental investment by the selecting parents (for example, by differential infanticide).

We do not argue that all cases of abortion constitute mate choice. Rather, we think that in the many possible circumstances under which control over prefertilization events is limited, females will be selected to increase control over reproductive commitments through postfertilization (and postunion) evaluations of the quality of male gametic contribution. (For similar interpretations of related phenomena traditionally viewed in terms of male-male competition, see Schwagmeyer, 1979; Labov, 1981.) When females use abortion as a mechanism of mate choice, males will be placed under counterselection pressure to acquire mechanisms that reduce the incidence of abortion of zygotes they father. Whether abortion should be viewed as female choice, parental manipulation, parent-offspring conflict, or possibly some other process, will depend on circumstance. For example, abortion of a deformed embryo by a faithful female paired for life to one male obviously does not constitute mate choice. On the other hand, we believe that differential abortion by a female with many eggs that were fertilized nearly simultaneously by multiple males can be profitably viewed as mate choice. In general, we hypothesize that mate choice can be invoked when a primary effect of post-"mating" destruction by a female (or female function of a hermaphrodite) of gametes, zygotes, or offspring is to foster adaptations by males (would-be "mates") to avoid such destruction.

By our definition of "mate," material investments provided to pollen that enhance the probability of offspring survival must be considered MI, rather than PI, if they function, as we hypothesize, to increase the probability of fertilization and to lower the probability of abortion. In allocating such resources, male plants must make final investment decisions earlier than many animals. That is, they must provide all investment prior to finding mates (before pollen is dispersed). Also, they cannot retrieve any wasted investment, such as that which fails to reach an appropriate receptive surface. And finally, they probably even lack mechanisms to prevent females from using their MI to enhance female RS even when their sperm is rejected (either pre- or postzygotically). This results in a situation analogous to cuckoldry, and it places males under selection pressure to limit their investment.

Sexual selection should not be an important force in species in which females typically have limited mating opportunities. Other circumstances may also diminish the importance of sexual selection. For example, in predictable environments, individuals might be able to gauge accurately their probability of reproductive success through male or female function. Individuals (such as protogynous hermaphrodites) could then "become" male only when their probable male RS is very high. While, admittedly, the evolution of such a pattern would be largely effected through sexual selection, the consequent reduction in the observed variance of male reproductive success would mean that sexual competition might not be evident in ecological time.

Another phenomenon that restricts the force of sexual selection is inbreeding. In a number of higher plants extensive inbreeding has evolved and selfing may have even have become the norm. Species that obligately self are not subject to the pressures of sexual selection (nor to kin selection in the sense discussed below) and are therefore outside the domain of this book. Many plants, of course, inbreed to a less extensive degree; in these sexual selection will be less important than in obligately outcrossed species, though probably not entirely absent. Selfed species may still engage in some processes that we believe may have resulted from sexual selection. That these processes may persist could be the result of historical constraints; for these plants, mechanisms that were previously used in "competitive" or "selective" processes are now merely the plants' way of getting something built or accomplished during development. The evolution of sexual selection mechanisms cannot be profitably explored in such species.

If we are to demonstrate that male-male competition and female choice are applicable to plants, we need to provide evidence that female RS is often limited more by resources for seed production than by number of matings obtained. Ideally, we would also show that male RS is limited more by number of mates than by resources. These issues are discussed, primarily with reference to seed plants, in the next chapter.

1.2 KIN SELECTION, PARENT-OFFSPRING CONFLICT, AND PARENTAL MANIPULATION

Hamilton (1963, 1964a,b) defined *r*, the coefficient of relatedness, as the probability that a gene possessed by an individual exists in a certain relative due to common descent (see also Wright, 1922; Malécot, 1948; Li, 1955). Numerous authors have employed variations of this definition (e.g., Starr, 1979); a commonly encountered definition of *r* is the fraction of genes shared by two individuals through common descent (West Eberhard, 1975). But this latter definition loses accuracy when there are asymmetries in relatedness between two individuals. Asymmetry occurs when individuals vary in ploidy (Crozier, 1970) and may occur in diploids when inbreeding (Flesness, 1978) and/or varying levels of homozygosity are present in a population.

A central concept in kin selection theory is "altruism,"

which refers to aid given to a relative at some cost to oneself. Hamilton (1964a,b) argued that the tendency of individual *A* to display altruism toward *B* should be proportional to *A*'s relatedness to *B* ($r_{A(B)}$). Crozier (1970; also Crozier and Pamilo, 1980) pointed out that *A*'s altruistic tendencies should be determined by *B*'s relatedness to *A* ($r_{B(A)}$), rather than the reverse. To appreciate the difference, imagine two partial sibs of a diploid species. These sibs share the same mother but have different fathers. Let the father of one be a relative of the mother (resulting in an "inbred" sib, *I*), while the father of the other is a nonrelative (resulting in an "outbred" sib, *O*). The relatedness of the inbred sib to the outbred one ($r_{I(O)}$) is greater than is true vice versa ($r_{O(I)}$). *I*'s offspring will, consequently, transmit proportionately more of *O*'s genes than will *O*'s offspring transmit of *I*'s genes. Crozier's refinement, that the tendency of one individual to behave altruistically toward another individual depends on the potential recipient's relatedness to the donor individual, only affects interpretation of results when asymmetries in relatedness exist. These asymmetries are the rule in the Hymenoptera, where kin selection has been most studied, and they also occur in some of the plant relationships discussed herein.

Individuals are selected to apportion energy strategically to maximize their "inclusive" fitness (Hamilton, 1964a,b). Inclusive fitness refers to the gene or offspring equivalents of an individual, produced by the individual and by relatives that are not direct descendants. For example, if an individual has ten offspring as well as ten sibs, each having ten offspring of their own, its inclusive fitness is based on sixty "offspring equivalents" or thirty "gene equivalents" in a diploid organism (ignoring other relatives' contributions). This obtains because an individual is half as related to its nieces and nephews as it is to its own offspring.

One theoretical issue concerns the evolutionary potential for conflict of interest between parents and their offspring.

Trivers (1974) argued that conflict occurs routinely between parent and offspring over the amount of parental investment or resources allocated. When the inclusive fitness of an offspring suffers from parental allocation of energy to other offspring, the offspring is selected to behave in ways that enhance the amount of resources it receives at the expense of sibs reared either simultaneously or subsequently. According to Trivers, three factors determine the optimal allocation of any parental resource from the perspective of a particular offspring: (1) the benefit it will receive from the resource, (2) the cost to the sibling(s) not receiving the resource, and (3) the degree of relatedness of the sibling(s) to the offspring in question. If, for example, a sib is related to a particular offspring by a coefficient of relatedness of 0.5, both that offspring and its parents would gain fitness if resources were allocated to the sib when the costs to the sib were greater than twice the benefits the offspring would receive if the resources were allocated to the focal offspring alone. If the cost-benefit ratio is less than one, again both parents and the offspring in question will gain if resources are allocated to the offspring instead of to its sib. Conflicts between parents and offspring occur when the cost-benefit ratio varies between 1 and 2. In these cases the parents' fitness is maximized by allocating some resources to the sib, but the offspring's fitness is maximized by having all resources allocated to it alone. Selection pressure on the offspring favors behaviors that signal that the offspring "needs" a greater share of resources. Pressure on the parents will favor the detection of deceit and measures to suppress or counter it. By this reasoning continuous conflict may occur in evolutionary time.

Hartung (1977) has pointed out that the potential for parent-offspring conflict is minimal when all offspring are full sibs, that is, when pair bonds are monogamous and partners are faithful. This occurs because an offspring is just as closely related to its own siblings as to its offspring. Thus an offspring that somehow causes its parents to increase their parental investment in it alone would merely increase its ability to manufacture offspring (which are related to it by 0.5) at the expense of its parents' ability to produce offspring that are full sibs (also related by 0.5). Actually, parent-offspring conflict could still occur, but the conditions are quite restrictive, requiring very specific patterns of resource distribution. Suppose, for example, an individual could substantially increase, but not quite double, its reproductive value if it could sequester the resources its parents "intended" for a yet-unborn sibling. Assuming that the sibling would have had the same reproductive expectation as the focal offspring had investment been allocated as the parents intended (e.g., with sufficient resources such that each could expect to produce four offspring), then the fitness of the parents would suffer by the reallocation of all resources to the single offspring, but the inclusive fitness of the focal offspring *could* increase (if it managed to produce seven, but not six, offspring instead of four). As Hartung noted, it is generally difficult to imagine finely tuned circumstances that would lead to such a net benefit for selfish offspring. Hartung (1980) later modified his position: "It is now clear to me that if, as Trivers stated, 'the benefit/cost ratio of a parental act changes continuously from some large number to some very small number near zero,' $1 < B/C < 2$ is a ratio that must be traversed during the normal course of parent-offspring relationships." However, the duration over which this condition holds (for example, when the parent is in the process of stopping "aid" to an offspring), will greatly affect the potential for significant conflict.

Alexander (1974) proposed the parental manipulation hypothesis, which argues that in evolutionary time parent-offspring conflict will be resolved in favor of parents. In his view this will result because individuals that are manipulated by their offspring will have lower inclusive fitness than those that control their offspring. An individual that successfully deceives its parents and thereby lowers their fitness will tend to produce deceitful offspring that in turn will lower its fitness. Parents that retain control over offspring obtain greater fitness; therefore controlling phenotypes will predominate in evolutionary time, and parents will manipulate their offspring to enhance their own fitness.

Empirical evidence for the alternative hypotheses of parent-offspring conflict and parental manipulation is equivocal, since it often can be interpreted to support either one. For example, starvation of some offspring will result if parents exert control over the amount of resources allocated to broods or to specific individuals; they may even make "extra" or secondary young to be routinely sacrificed unless the primary young die (Alexander, 1974). On the other hand, differential allocation of resources can be viewed as resulting from sibling rivalry, which may lower parental fitness and over which parents may lack control (Trivers, 1974). The occurrence of sibling rivalry does not necessarily imply conflict between parents and offspring, however, since competition among offspring may be advantageous to manipulative parents by providing evidence of relative offspring quality. The social organization of the eusocial Hymenoptera, which are typified by caste structures in which most females spend their lives contributing to colony growth without reproducing themselves, has been interpreted as resulting from control imposed by queens (West Eberhard, 1975; Alexander and Sherman, 1977) and from kin selection through parent-offspring conflict (West Eberhard, 1975; Trivers and Hare, 1976).

Complicating this debate is a sometimes acknowledged confusion (e.g., Craig, 1979; Crozier, 1979) over appropriate measures of *r*, the coefficient of relatedness. Some of this confusion may result from a blurring of the concepts of gene and individual (= genotypic) selection. For example, Crozier (1970) mentioned as a conceptual difficulty the position of a genotype "*CD*" (where *C* and *D* represent haploid genomes, not alleles at a locus), in evaluating the relative worth of an identical sib "*CD*" versus a homozygous "*CC*" individual as recipients of altruism. From the genotypic perspective the relationships (r) of individuals *CD* and *CC*, as well as a homozygous "*DD*" individual, to *CD* are all unity, even though one genotype is identical to *CD*, while the others are not. *CC, CD,* and *DD* individuals will transmit, on average, the same number of genes derived from a *CD* ancestor. Thus *CD* should be equally disposed to investing in the three genotypes. From the point of view of any of the *D* genes, however, it would be most advantageous to invest in *DD*, whereas from the perspective of any of the *C* genes, *CC* should receive the investment. Investing in *CD* represents a compromise. This consideration suggests that gene-level selection creates the potential for conflict within the genome, a view promulgated by Dawkins (1976). To the extent that this conflict lowers the fitness of an individual bearing conflicting genes, genes may be viewed as parasites of their host.

Another example of the blurring of individual and gene selection is the widespread use of gene frequency models borrowed from population genetics to investigate hypotheses formulated at the individual level. Several quantitative models have been developed at the gene level to explore the potential for parent-offspring conflict, but the bulk of these actually deal with sibling competition (e.g., MacNair and Parker, 1978, 1979; Parker and MacNair, 1978), which may or may not (Alexander, 1974; Hartung, 1977) imply parent-offspring conflict. These models support verbal statements that the potential for sibling competition appears greatest among offspring receiving simultaneous rather than sequential investment (Metcalf *et al.*, 1979; Parker and MacNair, 1979; see also Alexander, 1974) and among those less than maximally related because of multipaternity (MacNair and Parker, 1978, 1979; Metcalf *et al.*, 1979; Parker and MacNair, 1979; see also Hamilton, 1964a,b; Williams, 1975). In models dealing more directly with parent-offspring interactions, whether parents or offspring win depends heavily on initial assumptions (Stamps *et al.*, 1978; Parker and MacNair, 1979). Gene selection models that examine the potential for parent-offspring conflict and parental manipulation may come to quite different conclusions from those of simulations employing breeding "individuals" (e.g., Charlesworth, 1978; Craig, 1979). These different results may not be surprising since the unit of analysis differs.

It seems likely that an individual that could suppress gene conflict (making genes "cooperate") would reproduce more abundantly than individuals with "warring" genes and that individual-level control would evolve since genes cannot reproduce if their hosts do not (e.g., Gould, 1977a). This reasoning is similar to that of Alexander's parental manipulation hypothesis (1974; see also Alexander and Borgia, 1978). One could, like Dawkins (1976: 146-147), flip the coin and point out that an individual cannot reproduce without its genes. This approach, however, misses a built-in asymmetry: a gene that prevents its host from reproducing hits an evolutionary dead end, whereas an individual can reproduce and, theoretically, maximize its fitness without reproducing both alleles of all its genes. If individuals were not able to suppress their warring genes, it seems unlikely that living systems could sustain (and enhance over evolutionary time) their highly negentropic characteristics (see also Alexander, 1979). Thus, when there is conflict between individual and gene selection, selection at the level of the individual should tend to predominate. In some respects this situation parallels the conflict between group and individual selection. A gene that follows the dictates of gene selection in opposition to the individual may doom its descendants in much the same way an individual that adheres to group interests is fated when opposed by individual selection (Williams, 1966; Alexander and Borgia, 1978). In a sense, then, it may even be reasonable to view genes as "group" selected (see also Hartung, 1981). It is unwise at this point to conclude that selection is effective at only one level, however. Genes do engage in certain activities (for example, replication along a chromosome and meiotic drive) that indicate that competition within the genome may occur (Crow, 1979; Doolittle and Sapienza, 1980; Eberhard, 1980; Orgel and Crick, 1980; Maynard Smith quoted by Lewin, 1981; Cosmides and Tooby, 1981). To the extent that such competition does occur, intragenome disagreements over investing in kin should be minimized when all of the genes of the donor (or "altruist") are contained in the recipient.

Allowing for the possibility that gene and individual selection can conflict, it appears that the tendency to confuse or even equate the two (Dawkins, 1978; but see Alexander, 1979) is misguided. It therefore becomes important to ask how one might measure the potential conflict between gene and individual selection. To do this, it is first necessary to approximate the internal conflict within an individual regarding investment decisions. Recall that $r_{B(A)}$ is the measure by which *individual A* evaluates the relative value of investing in B. By contrast, $r_{B(A)}$ reflects the relative conflict within A's genome over investing in B. This holds because if all (or none) of the genes of the potential altruist are contained in the potential recipient, there will be less "conflict" among the genes of the altruist than at intermediate levels of genic relatedness. By this measure, for example, the donor genotype *CD* has a genic relatedness (1) to an identical sib *CD* that is twice its relatedness (0.5) to homozygous *CC*. There will, therefore, be less conflict within the genome if investment is placed in a *CD* recipient. As noted earlier, such conflict does not occur from the individual standpoint in this case, since *CC* and *CD* genotypes are equally related to *CD*. Nevertheless, if gene level competition were significant, the *CD* individual could suppress it somewhat by favoring *CD* beneficiaries over *CC* or *DD* ones. (This would suggest, of course, that strong intragenome conflict is one factor favoring the evolution of asexual reproduction. The prevalence of sexual reproduction would seem to indicate that individuals are commonly successful in suppressing intragenome conflict.) Gene and genotypic selection conflict least when the two measures of relatedness are equal.

Some types of interactions that might be considered examples of parent-offspring conflict, offspring-offspring conflict, or parental (female) manipulation may be more profitably viewed as conflict between the parents or competition among potential mates. An example involves the conflict over abortion. A reproductive tactic commonly employed by higher plants is the partial abortion of multipaternal "broods," probably in order to ensure investment in high quality offspring. Any offspring that successfully avoided abortion would have higher fitness than those that did not. But it is likely that the real conflict occurs between the female and the father(s) of to-be-aborted zygotes, not between mothers and zygotes or among the zygotes per se. To see this, it is necessary to consider the genetic relationships among the parties involved.

A given female may produce a certain number of ovules, expecting to abort some number after fertilization, while saving the remainder (z) for subsequent investment. The average relatedness between any two randomly selected zygotes depends upon the number of mates a female has as well as the variance in number of matings with each male. Regardless of the number of males and their variance in mating number, the average relatedness of any unfertilized egg (or any female portion of a given genome) to every other unfertilized egg (or other female portions of all zygotes) is ½. The average relatedness of a nonaborted offspring to a to-be-aborted zygote is 1/4 in the absence of inbreeding; it is greater than 1/4 if some inbreeding occurs or if zygotes have a common father. An aborted offspring leaves behind it on the parent a total of $z/4$ gene equivalents. In large broods $z/4$ is so great that the consequences of abortion for the aborted embryo are minimal. Any small advantage to avoiding abortion that might accrue would be strongly offset *if such an attempt lowered the average fitness of sibs by even a small amount.* For example, when $z = 200$, there are fifty gene equivalents of a to-be-aborted embryo in the mother's brood if there is a different nonrelated father for each offspring. If the embryo could substitute itself for one of the nonaborted offspring, the gene equivalents of the brood to it would increase to 50 3/4 (since its relatedness to itself is unity). The proportionate benefit an offspring accrues by avoiding abortion at the expense of a substituted sib declines as the number of mates decreases and/or z increases.

There are additional costs of embryo substitution, both to the mother and to the embryo itself. These costs include a depression in the mean genetic fitness of the brood (since the mother originally should have selected embryos of the highest quality), the expense to the mother in determining which embryo to substitute, and the cost of aborting an intended offspring in which she has placed substantial investment. This last cost should be nontrivial since the female should have opted to invest in more than 200 offspring if such investment would not have lowered the average fitness of the brood or would not have depleted her ability to invest in future broods (which may be equally related to present embryos). Since selection should optimize female investment in a brood, increased expenditure must lower the reproductive success of the mother and the inclusive fitness of offspring. These costs also should prevent doomed offspring from realizing their potential gain in inclusive fitness. If the brood is very small, the advantages of avoiding abortion may offset costs to to-be-aborted embryos, but the evolutionary potential for conflict between a large-brooded female and particular embryos is negligible.

By contrast, there is considerable potential for conflict between the male portion of the zygote and its mother. A pollen grain's relatedness to the mother's offspring is quite variable, depending on both the number of mates and their variance in mating success, and is therefore likely to be somewhat unpredictable. As the number of males increases, however, the relatedness between the male complement in an embryo and "sibs" approaches 0 (assuming no inbreeding). Therefore, the value of attempting to prevent abortion becomes very large and may even be enhanced if such action lowers the fitness of other zygotes.

In sum, the effects of a rare allele that confers increased resistance to abortion (without complicating pleiotropic effects and regardless of the "quality" of the genome in which it is found) differ depending on whether it is located in a maternal or paternal genome. Within the genome of a large-brooded female it would likely result in conflict between individual and gene selection rather than between parent and offspring, because if the allele spread, it would do so at the expense of the inclusive fitness of the brood as well as the reproductive success of the mother. Within the paternal genome such an allele would tend to increase the reproductive success of the father and the inclusive fitness of the father's "brood." This suggests, then, that selection would tend to suppress the effects of such an allele in maternal gametes and in the maternal portion of early zygotes and enhance their effects in paternal gametes and in the paternal portion of zygotes. Thus sex-limited expression of the allele would tend to evolve.

It is reasonable to view this situation as one of conflict between the parents rather than between parent and offspring because the interests of the female portion of the zygote should be coincident with those of its mother, while those of the male portion should not. In this view the practice of abortion represents postzygotic mate selection by females. Males that evolve mechanisms that lower the probability of abortion of their offspring will be strongly selected for, as will females that successfully counter such attempts.

The question then becomes which sex will win this evolutionary struggle. Alexander (1974) suggested that conflicts between parents are resolved in favor of the sex that invests more (usually female). Evolutionarily, this sex may have more at stake, and hence selection pressure may be stronger. At a proximate level this sex should usually possess greater control over the fate of offspring. While it may often be the case that circumstances favor maternal control, there may be instances in which fathers are able to counter female control. Since male gametes have the potential to evolve faster than do female gametes (because they are much more abundant: see Parker *et al.*, 1972), males may have an advantage disproportionate to their investment. Also, it is unlikely that females universally have more at stake in every offspring: if, for example, males have to commit much energy to effect a single fertilization, the resulting zygote is quite important. Furthermore, if male plants are vulnerable to having their MI used for rearing other males' offspring, as suggested above, the intensity of selection on them to avoid this outcome is considerable. Thus evolutionary conflict between males and females over abortion seems likely to occur and should be particularly intense when males contribute MI.

CHAPTER TWO

Limitations on Reproductive Success

FEMALES

True reproductive success is measured by the numbers of offspring that survive to reproduce themselves, and their relative success. We lack any measure of this for plants and for most animals, however. The only available measure of female RS in plants is seed production; the question then becomes, "What limits seed production in plants?"

Two major factors that may limit seed production are consumable resources (e.g., energy, minerals, water) and pollen availability. The sociobiological argument about mate competition and mate preference is predicated on the limitation of RS of one sex (usually female) by "food" resources and the limitation of RS of the other sex (male) by the number of fertilizations obtained. Accordingly, it is fundamental to provide evidence that such is frequently the case. Of course, we cannot be sure that present circumstances reflect the conditions under which a species evolved to its current state, but present conditions are our only source of information.

We do not claim that limitation of RS of either sex is always an either-or proposition. In the short term both resource and pollen limitation may contribute to determination of fruit production, as Wyatt (1981) has argued, although the experiments cited below suggest that one or the other factor may also work alone. Rather, we suggest that if resource limitation of female RS contributes significantly to a female's lifetime fecundity, the conditions are established for mate choice by females. This is not the case when the main constraint is pollen supply.

Evidence for resource limitation is provided by four kinds of data: (1) the effect on seed production of addition of mineral fertilizer, (2) the effect of competitive release, (3) the effects of defoliation, tapping of xylem tissues, or decrease of soil nutrients, and (4) reduced vegetative growth or subsequent fruiting of individuals as a consequence of reproduction.

It is noteworthy that genotypic differences exist in seed-producing ability (e.g., Croker, 1964, in Brazeau and Veilleux, 1976; Brown, 1971; Grisez, 1975; McLemore, 1975; Dorman, 1976; Varnell, 1976; Orr-Ewing, 1977), in cone abortion rates (Katsuta, 1971; White *et al.*, 1977), and in response to fertilizer application (Shoulders, 1967; Schultz, 1971; Jahromi *et al.*, 1976). These observations suggest that individuals in typically hermaphroditic species have inherent differences in their ability to function as females, and some may function primarily as males (Horovitz and Harding, 1972b).

Many studies, mostly from the forestry literature, provide evidence that the addition of mineral fertilizers commonly increases seed production (although there are some exceptions; also, results from those studies conducted in plantations may not reflect natural conditions). The summary in Table 1A indicates that resources (here, minerals) are frequently limiting to female RS. Removal of neighboring competitors is associated with increased seed production (Table 1B), and resource depletion often reduces female RS (Table 1C). A number of studies have shown that production of fruit is correlated with reduced vegetative growth or reduced fruiting in following year(s) as a result of internal allocational shifts of resources by the fruiting individual (Table 1D).

There is some evidence that levels of seed production

TABLE 1. Evidence for resource limitation of seed production[a]

Species	Result	References
	A. Effects of addition of mineral fertilizers (not all applications equally effective).	
Abies balsamea	increased seed size	Sheedy, 1974
Acer saccharum	increased seed numbers, weight, and % good seed	Chandler, 1938
Asclepias syriaca	increased pod production	Willson & Price, 1980
A. verticillata	increased pod production	Willson & Price, 1980
Chamaesyce hirta	increased reproductive effort	Snell & Burch, 1975
Coix Ma-yuen	increased seed production	Kawano & Hayashi, 1977
Cryptomeria japonica	increased seed production	Matthews, 1963
Fagus grandifolia	increased number sound fruit	Chandler, 1938
F. sylvatica	increased number sound fruit	LeTacon & Oswald, 1977
Hevea brasiliensis	increased fruit production	Matthews, 1963
Larix leptolepis	increased seed production	Brazeau & Veilleux, 1976; Matthews, 1963
L. gmelinii	increased seed quality	Brazeau & Veilleux, 1976
Leptospermum scoparium	increased seed production	Primack & Lloyd, 1980
Picea abies	variable effects (+, −, 0)	Chalupka, 1976; Maalkonen, 1971
P. glauca	increased seed production	Matthews, 1963
P. mariana	increased cone production	Brazeau & Veilleux, 1976
Pinus densiflora	increased cone & seed production	Brazeau & Veilleux, 1976
P. echinata	increased number of sound seeds	Dorman, 1976; Brazeau & Veilleux, 1976
P. elliottii	increased seed weight and cone production, genotypic differences	Matthews, 1970; Shoulders, 1968, 1973; Jahromi *et al.*, 1976; Dorman, 1976; Schultz, 1971; Brazeau & Veilleux, 1976
P. lambertiana	increased cone production	Brazeau & Veilleux, 1976
P. monticola	variable results	Bingham *et al.*, 1972; Brazeau & Veilleux, 1976
P. palustris	increased cone production; individual differences	White *et al.*, 1977; Allen, 1953; Brazeau & Veilleux, 1976; Snyder *et al.*, 1977; Shoulders, 1967, 1968; Croker, 1964; Dorman, 1976; McLemore, 1975
P. radiata	increased seed production	Matthews, 1963

TABLE 1. (con't)

Species	Result	References
P. resinosa	increased cone & seed production, esp. in older trees	Brazeau & Veilleux, 1976; Matthews, 1970; Fowler & Lester, 1970; Day *et al.*, 1972
P. strobus	increased seed production, esp. of older trees	Matthews, 1963; Brazeau & Veilleux, 1976
P. sylvestris	increased seed production (see text)	Sarvas, 1962; Brazeau & Veilleux, 1976
P. taeda	increased cone production of young but not older trees	Wenger, 1953
Pseudotsuga spp.	increased cone production in some years	Steinbrenner *et al.*, 1960; Silen, 1978; Ebell, 1972; Ebell & McMullan, 1970; Brazeau & Veilleux, 1976
Quercus alba	increased seed number and seedling vigor	Detwiler, 1943
Q. ilicifolia	increased fruit production	Wolgast & Stout, 1977
	B. Effects of competitive release[b]	
Araucaria cunninghamii	increased cone production	Florence & McWilliam, 1956
Diamorpha smallii	increased seed production	Clay & Shaw, 1981
Larix leptolepis	increased seed production	Brazeau & Veilleux, 1976
Melampyrum lineare	increased number of seed capsules	Cantlon, 1970
Pinus densiflora	increased seed production	Brazeau & Veilleux, 1976
P. echinata	increased seed production	Yocum, 1971; Dorman, 1976
P. elliottii	increased cone production	Dorman, 1976; Florence & McWilliam, 1956; Shoulders, 1968; Matthews, 1963
P. palustris	increased cone production	Dorman, 1976; Allen, 1953; Shoulders, 1968
P. radiata	increased seed production	Brazeau & Veilleux, 1976
P. resinosa	increased cone production	Fowler & Lester, 1970; Brazeau & Veilleux, 1976
P. taeda	increased cone & seed production, esp. in some individuals and in some years	Wenger, 1954; Pomeroy & Korstian, 1949; Grano, 1957; Dorman, 1976; Allen & Trousdell, 1961; Florence & McWilliam, 1956
Pseudotsuga menziesii	increased cone production	Silen, 1978
Quercus spp.	increased fruit production	Brazeau & Veilleux, 1976

TABLE 1. (con't)

Species	Result	References
Ranunculus bulbosus	increased seed production	Salisbury, 1942
Saxifraga tridactylis	increased seed production	Salisbury, 1942
Vulpia fasciculata	increased seed production	Watkinson & Harper, 1978
C. Effects of defoliation, tapping, decreased soil nutrients on reproduction		
Abies concolor	decreased cone production	Fowells & Schubert, 1956
Abutilon theophrastii	decreased seed production at high density	Lee & Bazzaz, 1980
Ambrosia artemisiifolia	decreased seed production and survivorship	Reed & Stephenson, 1972
Arctium minus	decreased seed production and seed weight	Reed & Stephenson, 1972, 1973
Asarum caudatum	decreased seed number	Cates, 1975
Asclepias syriaca	decreased pod production	Willson & Price, 1980
Catalpa speciosa	decreased fruit production	Stephenson, 1980
Cordia macrostachya	decreased flower & fruit production	Callan, 1948; Simmonds, 1951
Fagus sylvatica	decreased seed production	Matthews, 1963
Gymnocladus dioicus	decreased seed production	Janzen, 1976
Pinus roxburghii	variable results	Lohani & Kureel, 1973
P. sylvestris	decreased seed production	Plancke, 1922, in Lyons, 1956
Prunus serotina	no flower production	Farmer & Hall, 1971
Rosa nutkana	decreased seed production	Myers, 1981
Rumex crispus	decreased seed weight sometimes	Bentley *et al.*, 1980; Maun and Cavers, 1971
R. obtusifolius	decreased seed number & seed weight	Bentley *et al.*, 1980
Solidago canadensis	galls decrease flower & fruit production	Hartnett & Abrahamson, 1979
Vaccinium corymbosum	decrease fruit production	Aalders *et al.*, 1969
D. Effects of reproduction on allocation within individuals[c]		
Abies grandis	reduced vegetative growth	Eis *et al.*, 1965
A. balsamea	reduced vegetative growth	Matthews, 1963; Morris, 1951; Powell, 1977a,b
Asclepias quadrifolia	reduced ability to flower & fruit in following years	Chaplin & Walker MS
Betula spp.	reduced vegetative growth	Gross, 1972
Chamaelirium luteum	reduced frequency of flowering in females	Meagher & Antonovics, 1982
Elaeis guineensis	reduced production of female flowers and fruit	Hartley, 1970

Table 1. (con't)

Species	Result	References
Fagus sylvatica	reduced vegetative growth	Matthews, 1963; Murneek, 1939
Picea abies	reduced vegetative growth	Matthews, 1963; Buyak, 1975
Pinus elliottii	reduced vegetative growth	Shoulders, 1967, 1968, 1973; Smith & Stanley, 1969
P. monticola	reduced vegetative growth	Eis *et al.*, 1965
P. palustris	reduced vegetative growth	Shoulders, 1968; Pessin, 1934
P. ponderosa	variable effects	Daubenmire, 1960
P. radiata	reduced vegetative growth	Matthews, 1963
P. resinosa	insect damage to flowers & fruits increases ability to reproduce the following year	Mattson, 1978
P. sylvestris	some variable effects	Chalupka *et al.*, 1976
P. taeda	reduced vegetative growth	Wenger, 1957
Prunus persiae	reduced vegetative growth	Daubenmire, 1960
Pseudotsuga sp.	reduced vegetative growth	Eis *et al.*, 1965; Ebell, 1971; Silen, 1978; Tappeiner, 1969
Pyrus (*Malus*) sp.	fruit set increases if growing shoots are pruned	Priestly, 1970, in Gardner, 1977; Luckwill, 1970
Quercus spp.	reduced vegetative growth	Murneek, 1939
many spp.	reduced vegetative growth	Kozlowski, 1971; Kozlowski & Keller, 1966
certain crops	reduced vegetative growth	Sinclair & deWit, 1975; Murneek, 1926; Leonard, 1962

[a] See also Lloyd, 1980a.
[b] See also Matthews, 1963.
[c] See also Harper & White, 1974.

are often determined early by the female parent (Lyons, 1956). For example, in *Catalpa speciosa* inflorescences become functionally male after a specific number of constituent flowers have been pollinated (Stephenson, 1979). Sometimes the size of the seed crop appears to be determined before the identity of incoming pollen is determined. The number of seeds per cone in *Pinus sylvestris* is reported to be specific to each mother tree regardless of self/cross ratios (Brown, 1970; Plym Forshell, 1974). In *Pinus elliottii* parents that produce many seeds when cross-pollinated generally produce many seeds when selfed (Snyder, 1968). In *Acer saccharum* selfing ability and cross-compatibility are positively correlated: the same trees that are most cross-compatible are also the most self-compatible; furthermore, the degree of self-fertility varies under different growing conditions (Gabriel, 1967). Annual as well as individual differences in degree of self-sterility are reported for several species (Blinkenberg *et al.*, 1958; Whitehouse, 1959; Sorensen, 1971; Clausen, 1973). Hill-Cottingham and Williams (1967) suggested that increased supplies of nitrogen during the appropriate season may prolong ovule life (in apples), allowing more time for the slower-growing self pollen to reach the ovules. This might enhance the frequency of selfing; at least it widens the range of pollen types for the female to choose from. These results suggest that seed production may often be determined less by the quality of the arriving pollen than by available resources or prevailing environmental conditions and that the degree of selectivity exercised by females (e.g., against selfing) can vary according to how many seeds are to be set as well as the proportions of different pollen types.

Differential abortion of few-seeded fruits (in *Pinus sylvestris* and *Juniperus*, see Sarvas, 1962; in domestic apples, see Luckwill, 1970; in *Hybanthus prunifolius*, see Augspurger, 1980, 1981b; in *Catalpa speciosa*, see Stephenson, 1980) and of herbivore-damaged seeds (e.g., Mattson, 1978) likewise indicates that the parent tree is allocating its resources to support the maximum number of seeds per unit of investment in fruit. However, the extensive development of parthenocarpy (development of fruits with no viable seeds) in many species (e.g., for angiosperms, see Nielsen and Schaffalitzky, 1953; Davis, 1966 (*Vitis*); Kriebel and Gabriel,

1969; Wilcox and Taft, 1969; Grundwag, 1975; for gymnosperms, see Sarvas, 1955; Fowler, 1965b; Mikkola, 1969; Hashizumi, 1973b; Plym Forshell, 1974; Lindgren, 1975; McLemore, 1977; Silen, 1978) might suggest either that fruit development is not resource limited and that parthenocarpy is merely a "mistake" resulting from induction by foreign pollen or that there is development error. Photosynthesis by sterile fruits themselves could reduce their costs to the parent (Bazzaz *et al.*, 1979), although perhaps the effort of making sterile fruits might be better directed toward making other kinds of tissue, such as leaves. Perhaps sterile "fruits" serve some important but as yet undiscovered function. One untested possibility suggested by D. A. Goldman (pers. comm.) is that sterile but well-developed fruits dilute the depredations of certain would-be seed predators (*if* they cannot detect externally the absence of seeds).

Altogether, then, numerous studies provide evidence of resource limitation of female RS. Exceptions seem to be few; that they exist at all is not surprising but should prompt closer examination of the conditions under which resources are not limiting to female function.

On the other hand, seed production by a female is sometimes limited by available pollen rather than "food" resources (Table 2, and see Denison and Franklin, 1975). Unfortunately, we have no idea if the introduction of the honeybee (*Apis mellifera*) to North America has reduced the frequency here of pollen limitation in insect-pollinated species. If controlled pollination results in greater seed production than natural pollination, the usual interpretation is that pollen quantity is limiting. Ideally, the effects of experimentally increased seed production on future reproduction need to be monitored to determine that resource expenditure in one season has no repercussions on subsequent crops (Janzen *et al.*, 1980), but this is seldom done. Individuals of some perennial species that seem to

Table 2. Summary of experimental evidence for pollen limitation of seed production: controlled pollination results in greater seed set than natural pollination (+) or not (−).

Species	Results	Comments	References
Aesculus californica	−		Benseler, 1975
A. pavia	−		Bertin, 1982a
Agave chrysantha	−		S. Sutherland, pers. comm.
Arisaema triphyllum	+		Bierzychudek, 1981
Asimina triloba	+		Willson & Schemske, 1980
Brassavola nodosa	+		Schemske, 1980a
Campsis radicans	+		Bertin, 1982b
Catalpa speciosa	−		Stephenson, 1979
Cryptomeria japonica	+?	increased pollination of ovules; no data on seed production	Hashizumi, 1973a
Elaeis guineensis	+	only at very low frequencies of male flowers	Hartley, 1970
Encyclia cordigera	+	(see text)	Janzen *et al.*, 1980
Erythronium albidum	+		Schemske *et al.*, 1978
Frasera caroliniensis	−		Threadgill *et al.*, 1981
Houstonia caerulea	+		Wyatt & Hellwig, 1979
Leptospermum scoparium	−		Primack & Lloyd, 1980
Liriodendron tulipifera	+		Wilcox & Taft, 1969
Mangifera indica	±	conflicting reports	Singh, 1960
Oenothera fruticosa	−		Silander & Primack, 1978
Paeonia californica	−		Schlising, 1976
Penstemon digitalis	−		Lee & Willson MS
Phlox divaricata	+		Willson *et al.*, 1979
Pinus elliottii	−		Dorman & Squillace, 1974
P. monticola	−		Bingham *et al.*, 1972
P. palustris	(+)	only rarely	Snyder *et al.*, 1977
P. ponderosa	−?	conflicting statements	Sorenson, 1970; Wang, 1970, 1977
P. sylvestris	variable		Sarvas, 1962
P. virginiana	−		Kellison & Zobel, 1974
Pistacia sp.	+?	some exceptions	Grundwag, 1975
Podophyllum peltatum	+		Swanson & Sohmer, 1976
Polygonum commutatum	variable	habitat differences	Lee & Willson MS
Prosopis spp.	±	one species, the remainder not	Solbrig & Cantino, 1975
Pseudotsuga sp.	variable		Silen, 1978; Daniels, 1978
Tectona grandis	+		Hedegart, 1973, 1976
Trientalis europea	+		Hiirsalmi, 1969
Vanilla fragrans	+		Purseglove, 1972
Yucca whipplei	−		Udovic, 1981

be pollen limited (e.g., *Arisaema triphyllum* and certain orchids) exhibit signs of resource limitation if they do get pollinated, and they reduce plant size or possibly shift to less costly male function in the following year (T. Gilmore, pers. comm.; Policansky, 1981; Bierzychudek 1982). In such plants both pollen and resource limitation seem to be present. Furthermore, experimental pollinations on one portion of a plant may raise seed set there but result in lower seed set elsewhere on the plant, implying fundamental resource limitation, so the whole plant must be monitored (S. Sutherland, pers comm.). If controlled pollination fails to elevate seed production above natural levels, the results are ambiguous: pollen may indeed not be limiting, *or* technical problems may have interfered with the success of controlled pollination. The latter is a common interpretation in the forestry literature (e.g., Snyder and Squillace, 1966; Hall and Brown, 1976), but the evidence for resource limitation (Table 1) suggests that negative results may often warrant the first interpretation: that of unlimiting pollen supplies. Extensive pollen dilution (from 100% to 10%) in experimental pollinations of *Pinus sylvestris* has little effect on seed set, although this species is often claimed to be pollen limited (e.g., Sarvas, 1962).

An increase of seed yield in *Pinus sylvestris* associated with greater site fertility was interpreted by Sarvas (1962) as being due to taller trees, increased pollen production, and better levels of pollination. It is not clear, however, whether greater seed production was indeed simply a result of greater pollen abundance or if it was the result of some other factor, specifically resources. In some populations pollen is indeed limiting to seed production in certain parts of the season (Schemske, 1977; Zimmerman, 1980) or under particular circumstances (Schemske, 1980b; Howell and Roth, 1981). But in general this condition seems to be much less prevalent than one of resource limitation (orchids may be exceptional: see Dodson, 1962). Information for one nonseed plant (*Equisetum* sp.) indicates that production of sporophytic offspring by gametophytes is not limited by fertilizations but rather by resource conditions influencing females (Duckett and Duckett, 1980). If large inflorescences have evolved to improve attraction of pollinators, as suggested by Schaffer and Schaffer (1977, 1979) for semelparous *Agave* species, then life history and eventual inflorescence size may be resource limited. In such a case pollen limitation and resource limitation are closely intertwined in evolutionary time. S. Sutherland (pers. comm.) has evidence that large agave inflorescences do not set proportionately more fruits than small ones, however; their role in pollinator attraction relative to female function and seed set is negligible.

Pollen *quality* is another problem. Inbreeding and outbreeding are often presented as dichotomous reproductive tactics, but is is more productive to envision a continuum of alternatives. Recently, several authors (Bateson, 1978; Lloyd, 1979; Price and Waser, 1979; Thiessen and Gregg, 1980; see also Wallace, 1958) have suggested that organisms may benefit from optimizing, rather than maximizing, outbreeding.

Inbreeding depression is common in many outcrossing species, typically appearing as lowered seed yield, germination success, and/or seedling growth and survival (Table 3 summarizes information from the forestry literature), and perhaps delayed age of first reproduction (Bingham, 1973). Thus inbreeding may have a marked influence on effective fecundity and eventual numbers of surviving offspring: it is customary to argue that inbreeding is therefore selectively disadvantageous.

At the other extreme, outbreeding depression may occur. Retarded growth of pollen tubes in *Betula verrucosa* and *B. pubescens* results not only from self-pollinations and interspecific crosses but also from some crosses between conspecific parents separated by considerable distance

TABLE 3. Summary of female fertility variations and effects of "inbreeding" in trees

	Individual variation in fertility	Cone yield	Number sound seeds	Germination success	Seedling growth; survival	References
GYMNOSPERMS						
Abies procera			–	0	–/0	Sorensen *et al.*, 1976
Larix dahurica				–		Franklin, 1970
L. decidua	+	0			–/+	Keiding, 1968; Franklin, 1970
L. europea				–		Franklin, 1970
L. leptolepis	+			+	–	Keiding, 1968; Franklin, 1970
L. sibirica				0		Franklin, 1970
Picea abies	+	–	– (usually	–	–/+	Eriksson *et al.*, 1973; Koski, 1973; Mergen *et al.*, 1965; Andersson, 1965; Andersson *et al.*, 1974; Franklin, 1970
P. asperata			–			Franklin, 1970
P. glauca	+		–		variable/–	Nienstaedt & Teich, 1972; Mergen *et al.*, 1965; Coles & Fowler, 1976; Franklin, 1970
P. jezoensis			–			Franklin, 1970
P. montigena			–			Franklin, 1970
P. omorika			0/+	+	–	Franklin, 1970; Koski, 1973; Fowler, 1965a; Sarvas, 1968
P. pungens	+		–			Franklin, 1970; Hanover, 1975
P. sitchensis			–			Franklin, 1970
Pinus banksiana	+	+	–	0	–/+	Fowler, 1965c; Franklin, 1970
P. cembra			–			Sarvas, 1968
P. densiflora		–	–	–		Franklin, 1970

P. echinata		−				Franklin, 1970
P. elliottii	+	− (usually)	−	−	−/+	Dorman & Squillace, 1974; Krause & Squillace, 1964; Franklin, 1970; Snyder, 1968
P. griffithii		+	0			Franklin, 1970
P. jeffreyi				−	−	Franklin, 1970
P. monticola	+	−/0	−	−	−/+	Bingham, 1973; Bingham *et al.*, 1972; Franklin, 1970; Barnes, 1964; Barnes *et al.*, 1962; Squillace & Bingham, 1958
P. palustris		0	−			Snyder *et al.*, 1977; Franklin, 1970
P. parviflora		−	+			Franklin, 1970
P. peuce			+/−	−		Kellison & Zobel, 1974; Koski, 1973; Sarvas, 1968
P. ponderosa	+		−	−/0	−	Sorensen, 1970; Sorensen & Miles, 1974; Wang, 1970, 1977
P. radiata		+/0	−	−	variable	Franklin, 1970
P. resinosa		+	0	0/+	−/0	Fowler, 1965a; Fowler & Lester, 1970; Franklin, 1970
P. rigida			−	0		Ledig & Fryer, 1974
P. strobus		−	−	+	−	Fowler, 1965c; Franklin, 1970
P. sylvestris	+	+	−	−	−	Plym Forshell, 1974; Franklin, 1970; Koski, 1973
P. tabulaeformis		−				Franklin, 1970
P. taeda		+	−	−	−	Franklin, 1970, 1971; Dorman & Zobel, 1973
P. thunbergii		−	−	−	+	Franklin, 1970; Kellison & Zobel, 1974

TABLE 3. (con't)

	Individual variation in fertility	Cone yield	Number sound seeds	Germination success	Seedling growth; survival	References
Pseudotsuga menziesii	+	−/+	variable	−	−(usually)	Orr-Ewing, 1957, 1976; Franklin, 1970; Sorensen, 1971, 1973; Sorensen & Miles, 1974; Piesch & Stettler, 1971
ANGIOSPERMS						
Acer saccharum	+					Gabriel, 1967; Kriebel & Gabriel, 1969
Betula allegheniensis	+					Clausen, 1973
B. verrucosa (self-incompatible)	+					Hagman, 1971
B. pubescens (self-incompatible)	+					Hagman, 1971
Fagus sylvatica	+					Blinkenberg *et al.*, 1958; Neilsen & Schaffalitzky, 1953
Liriodendron tulipifera (self-incompatible)	+					Wilcox & Taft, 1969
Tectona grandis			+			Hedegart, 1973; Nielsen & Schaffalitzky, 1953

NOTE: In most cases authors were actually referring to selfing; for self-incompatible species, this is not the case. Code: inbreeding resulted in an increase of >5% (+), a decrease of <5% (−), less than 5% change (0), no data (blank). Two different signs separated by a slash indicate conflicting reports.

(Hagman, 1971, 1975). Price and Waser (1979) suggested for *Delphinium nelsonii* that pollen from sources at an intermediate distance from the recipient effected higher mean seed set per flower than pollen from either shorter or greater distances. A similar phenomenon has been reported for some members of the *Stylidium crassifolium* species complex (Banyard and James, 1979). Thus there may be an "outcrossing depression" similar to the inbreeding depression much more commonly studied (e.g., Morton *et al.*, 1956; Sorensen, 1969; Sved and Ayala, 1970; Franklin, 1972; Klekowski, 1972; and Table 3). The possibility of outbreeding depression suggests the existence of genetic differentiation of populations along ecological gradients or across ecological thresholds. Several empirical studies have indicated that such gradients or thresholds can result in local population differentiation in spite of considerable gene flow (e.g., Jain and Bradshaw, 1966; McNeilly, 1968). The existence of gradients should favor enhanced recognition of mate quality. Mathematical models indicate that local differentiation should occur only when gene flow is quite restricted (Rohlf and Schnell, 1971; Endler, 1973, 1977).

There may be evolutionary limits to selection against inbreeding. To the extent that neighboring individuals are relatives and yet the most likely source of nonself pollen (S. Wright, 1943, 1946; J. W. Wright, 1952; Bateman, 1947a, b,c, 1950; Langner, 1952, in Bannister, 1965; Erdtman, 1954; Sarvas, 1962, 1968; Whitehead, 1969; Gleaves, 1973; Levin and Kerster, 1974; Coles and Fowler, 1976; but see Lanner, 1966; Levin, 1981), it may be difficult to select for strict self-infertility, at least when it is based on the genetic constitution of the gametes. Furthermore, Wells (1979) and others (see Lloyd, 1979) have argued that self-pollination will evolve and persist in populations with a capacity for outcrossing because facultative selfers may enjoy greater success as pollen donors, spreading the selfing gene to outcrossers (but see also Charlesworth and Charlesworth, 1981).

A potential cost of self-incompatibility (and obligate outcrossing) arises especially when only a few loci or alleles determine rejection of pollen (Lewis, 1979). In these cases grains that are not related to the female but have the same allele(s) are rejected, regardless of the rest of their genetic makeup. Females that discriminate in this way reduce the range of possible fathers for their offspring and, if pollen limits female reproductive success, may also reduce their seed crop size.

Indeed, selfing may be quite advantageous in several ways (see also Lloyd, 1980b). Many botanists (e.g., Baker, 1955; Bannister, 1965) have suggested that selfing is advantageous in colonizing species because it reduces problems with mate-finding and may permit offspring that are thus similar to the parent to occupy and perhaps preempt space like that occupied by the parent. The success of the latter possibility depends, of course, on the size and variability of patches. If there are costs to the avoidance of selfing, then the ability to self can reduce those costs and permit allocation of resources to production of greater numbers of offspring (e.g., Lloyd, 1979), which is also advantageous in colonizing species (Willson, 1981). Small population size, the expense of dispersal, ecological and geographical marginality, and ecological patchiness likewise can select for the ability to self (or at least for tolerance of selfing and of lesser degrees of inbreeding) (e.g., Allard *et al.*, 1968; Antonovics, 1968, 1976; Williams, 1975; Vasek and Harding, 1976; Bengtsson, 1978). Unpredictable fluctuations in pollen availability may also favor selfing, a selection pressure that appears to be more intense in annuals than perennials (Stebbins, 1950; Baker, 1959; Lloyd, 1980b). An unexplored possibility for plants is that kin-selected adaptations may also favor inbreeding (Thiessen and Gregg, 1980). When selfed or inbred siblings grow as neighbors, competition among them has less influence on the fitness of the losers than it would among nonrelatives. The actual level of competition may be either more or less intense. If it is less, the sibs may be able to afford slower growth and reap whatever benefits may accrue (see also Sakai *et al.*, 1968), including greater fecundity (Nienstaedt and Teich, 1972) and perhaps reduced risks of wind-breakage (Cech *et al.*, 1976). If competition is greater, however—because of the similarity of the sibs' phenotypes—slower growth may represent a cost rather than a "luxury."

If selfing were strongly disadvantageous, we might expect higher levels of self-sterility than seem to be the case. (Even when outcrossing is obligatory, it is by no means assured that the sole advantage in all cases lies in outcrossing itself [Willson, 1979; Bawa, 1980].) In fact, plants have a variety of mechanisms that may allow them to adjust the frequency of self-fertilization. "Autotoxicity" may reduce the relatedness of neighboring plants (Smith, 1979), thus increasing the probability of being fertilized by nonrelatives. Male and female functions of hermaphroditic plants may be positioned strategically on different parts of a flower, inflorescence, or the plant itself (Proctor and Yeo, 1973; Maynard Smith, 1978; Faegri and van der Pijl, 1979). Conifers might vary the size of the pollen chamber, where incoming pollen is deposited, and thus vary the frequency of outcrossing (Sarvas, 1968; Stern and Roche, 1974) without preventing self-fertilization completely. Large pollen chambers may increase the probability of deposition of outcross pollen, especially at high pollen densities, because the chamber is less likely to be filled by self pollen. Changes in the area of an angiosperm stigma or in the numbers of flowers (some of which may be aborted) might also accomplish this end. Floral size may affect rates of cross-pollination, as in *Lycopersicon pimpinellifolium* (Rick *et al.*, 1978), *Gilia* spp. (Grant and Grant, 1965; Schoen, 1977), and others.

Differences in timing of male and female functions may also reduce inbreeding (Darwin, 1876; Cruden and Hermann-Parker, 1977; Faegri and van der Pijl, 1979; Thomson and Barrett, 1981): protandry and protogyny can reduce the likelihood of deposition and success of self pollen. Sarvas (1962) noted that female cones of *Pinus sylvestris* typically open between one-half and one day in advance of male cones; he suggested that this reduces selfing. Pollen shed by neighboring individuals tends to be well synchronized (see also Jones and Newell, 1946; Bramlett, 1973), and the timing appears to be affected by atmospheric conditions in ways that may maximize dispersal (Sarvas, 1962, 1968). It may be difficult for female organs to match exactly the timing of local pollen shed; when local synchrony occurs, it then would be more advantageous to become receptive in advance of pollen shed by neighbors (and remain receptive) rather than being late and receiving little pollen. Furthermore, females that anticipate pollen shed by synchronized neighboring individuals may receive pollen from more distant sources, thus reducing not only selfing but also inbreeding with relatives.

A mixed strategy of inbreeding plus outcrossing may combine the advantages of each (Solbrig, 1976), as discussed also by Capinera (1979) for seed size. It would be most interesting to know if mothers can allocate resources in such a way (by building different sizes of dispersal units, by positioning flowers, and so on) that outcrossed offspring are more likely to be long-distance dispersers while selfed or highly inbred young are likely to disperse locally. For instance, outcrossed seeds, from chasmogamous flowers, of *Impatiens biflora* are larger and typically borne more distally on the parent than are seeds produced by selfing in cleistogamous flowers (Schemske, 1978; Waller, 1980). All seeds are explosively discharged so their size should affect the distance they disperse. Whether the decreased dispersal distance of the larger seeds is countered by their distal position is not known. But outcrossed seeds of *Gymnarrhena micrantha* are wind dispersed, while seeds from cleistogamous flowers are subterranean, much larger, and not dispersed at all (Koller and Roth, 1964; Zeide, 1978).

We would expect variability among genotypes in the degree to which selfing is advantageous because of differing frequencies of recessive lethals and differing levels of genetic complementarity. Indeed, both the degree of self-fertility and the magnitude of its effects, including the level of inbreeding depression, are known to differ among individuals (Drayner, 1959; Whitehouse, 1959; Barnes, 1964; Gabriel, 1967; Valdeyron *et al.*, 1977; Free, 1970). Female outcrossing rates differ among genotypes of *Lupinus nanus*, not merely as a result of different levels of self-fertility (Horovitz and Harding, 1972a). Genotypic differences in percentages of outcrossed seeds are known in cotton (*Gossypium*) (Stephens, 1956). Performance as female parent was inversely correlated with success as male parent in horticultural *Freesia* (Sparnaaij *et al.*, 1968) and in *Campsis radicans* (Bertin, 1982c). Incompatibilities or differential acceptabilities are reported between certain individuals in several angiosperm species (*Acer saccharum*, see Gabriel, 1967; *Liriodendron tulipifera*, see Wilcox and Taft, 1969; *Borago officinalis*, see Crowe, 1971; *Medicago sativa*, see Barnes and Cleveland, 1963; *Betula* spp., see Hagman, 1971; and *Campsis radicans*, Bertin, 1982c), and it would not be surprising to find also in conifers (e.g., *Pinus monticola*, see Barnes *et al.*, 1962) that crosses between certain genotypes are less productive than others. Differential ability of males to fertilize is known for several species, and such differences often vary with the genotype of the maternal parent (Squillace and Bingham, 1958; Barnes and Cleveland, 1963; Pfahler, 1965, 1967; Sparnaaij *et al.*, 1968; Sari Gorla *et al.*, 1975). Genotype differences in pollen production and potential for pollen donation are inferred for cotton (Stephens, 1956). Furthermore, different genotypes may have inherently different rates of pollen tube growth, unrelated to compatibility or ratios of self and cross pollen (Jones,

1928; Barnes and Cleveland, 1963; Mulcahy, 1974b; Ottaviano *et al.*, 1975; Sari Gorla *et al.*, 1975; Wolff, 1975; Socias i Company *et al.*, 1976; Bookman MS), although it is often not known whether the effect is due to the genotype of the pollen itself or of its parent (Baker, 1975; Sari Gorla *et al.*, 1975). *Asclepias speciosa* pollen from different sources produced pollen tubes of different sizes; pollen donors characterized by larger pollen tubes often fathered more seeds per pod and had larger seedlings (at age one month) than those with shorter pollen tubes (Bookman MS). Similar results were found in *Campsis radicans* (Bertin, 1982c). Thus variation in competitive ability of pollen was associated with apparently higher reproductive success. These observations have important implications in that they demonstrate the existence of individual variation from which females may choose.

The balance between effort devoted to outcrossing and that to inbreeding should be affected by environmental factors. For example, if outcross pollen is in poor supply, the threshold for acceptance of self pollen should decrease. If self pollen greatly outnumbers cross pollen, it sometimes prevails over cross pollen even on parents that normally favor cross-pollination (Barnes *et al.*, 1962). Furthermore, the effects of inbreeding on seedling growth and survival may well vary with prevailing ecological conditions (Markerian and Olmo, 1959; Andersson *et al.*, 1974).

Even when all arriving pollen comes from the same source, there may be differences in quality (e.g., Mulcahy *et al.*, 1975; Mulcahy, 1979). When large amounts of self pollen were placed on the stigmas of several crop species, offspring variability was less than when few self pollen grains were used (similar results were obtained from forced crosses between strains of *Triticum vulgare*: see Ter-Avanesian, 1978). These results suggest that at low levels of pollen competition most or all pollen was acceptable but that at high levels only certain pollen succeeded. By presenting conditions that produced a "race" among pollen grains, females have, in effect, found a means of discrimination among pollen. They could therefore be said to be more discriminating when more pollen is available.

Furthermore, pollen competition is likely to be greater when the pollen tubes have to grow a long distance to reach the ovules than when the distance is short (Correns, 1928, in Mulcahy and Mulcahy, 1975). Pollen from a single donor of *Dianthus chinensis* was placed on the elongate stigma at different distances from the ovules. Seeds fathered by distally placed pollen germinated slightly but significantly faster and produced larger seedlings than those sired by pollen placed proximally. The variance of germination time was also less for distal pollen. Mulcahy and Mulcahy (1975) interpret these results as an indication of greater sorting of pollen (either by themselves or indirectly by the female) having the greater distance to grow. Because these experiments used pollen from a single donor, the results indicate considerable variation even for pollen from one individual; it would be interesting to compare results from multiple fathers as well.

The results of depositing different amounts of pollen on stigmas of *Petunia hybrida* are not so readily interpreted in this way. Pollen competition presumably increases with the amount of pollen on the stigma. Using pollen from one donor clone deposited in differing amounts on stigmas of another clone of *P. hybrida*, Mulcahy *et al.* (1975) could not show that seeds or seedlings resulting from fertilizations achieved under conditions of intense pollen competition were generally larger, faster developing, or more numerous; the only exception was found in ninety-two-day-old seedlings, when the initial advantages of those from lightly pollinated flowers were reversed. The authors interpret the results to mean that "maternal effects" outweigh competitive effects until day ninety-two; they may be correct, but the system warrants closer investigation.

It is clearly important to establish the general occurrence of pollen competition and the association of success in such competition with fitness of the ensuing sporophyte. Much of the work so far concerns crop plants and cultivars. Correlations between gametophytic (pollen) and sporophytic (seed and seedling) traits are documented for *Zea mays* (Mulcahy, 1971, 1974a; Ottaviano *et al.*, 1980) and may involve both nuclear and cytoplasmic factors (Pfahler, 1975). Associations of pollen and offspring traits are recorded for other domesticated genera such as cotton, wheat, beans, peas, and tomatoes (Jones, 1928; Mulcahy, 1974b). Tanksley *et al.* (1981) found overlapping gene expression in tomato pollen and sporophytic offspring; earlier, Schwartz (1971) reported similar overlap in *Zea mays*. When female *Drosophila melanogaster* were allowed to choose their mates (rather than being randomly paired), one component of offspring fitness was enhanced (Partridge, 1980). These results clearly indicate that male-male competition and female choice *might* be important to the resulting offspring. Mulcahy (pers. comm.) has indicated, however, that unpublished results for other species fail to exhibit these interesting correlations.

One other factor that affects female success needs brief discussion. Seed predators are sometimes able to make great inroads on seed crops and may demolish some crops entirely. Seed predation limits recruitment to populations of *Haplopappus squarrosus* (Louda, 1982). Virtually total destruction of seeds by predators, especially in poor seed years, is frequently reported for conifers (e.g., Barner and Christiansen, 1962; Larson and Schubert, 1970; Mattson, 1971; DeBarr and Barber, 1975; Goyer and Nachod, 1976; Snyder *et al.*, 1977) and oaks (Downs and McQuilkin, 1944; see also Janzen, 1971a). These depredations are likely to be countered evolutionarily by changes in resource allocation patterns within individuals, however. Thus, in the long run, effects of predation should influence the resource budget (but be unrelated to pollen limitation) and have no direct effect on the dichotomy at issue here.

2.2 MALES

What limits male RS? In general, we expect that it is more often limited by the number of mates than by resources. But in this respect many plants may be more complicated than most animals in that most plants are hermaphroditic, producing both male and female gametes.

Ecologists generally expect parental expenditure on offspring of each sex to be equal, as was argued by Fisher (1958) for large, panmictic populations in which progeny sex ratio is a genetically determined character. It is sometimes thought that Fisherian conditions are applicable to the allocation, by hermaphroditic parent plants, to male and female gametophytes, as represented by pollen grains and ovules (e.g., Smith, 1981). If true for individuals, then pollen production is limited to the number of grains that can be produced by half of the total reproductive resource budget of the parent. It could then, perhaps, be claimed that pollen production is (indirectly) as resource limited as ovule and seed production. We do not support this contention, however, for two categories of reasons.

(1) Deviations from Fisherian conditions are expected. In dioecious populations, when reproduction occurs in patches colonized by relatively few individuals, the equilibrium sex ratio may be highly skewed toward female production (Hamilton, 1967; Taylor and Bulmer, 1980). Also, the equilibrium population sex ratio may be biased in favor of the sex that disperses more evenly and/or more widely (Bulmer and Taylor, 1980; see also Clark, 1978). We now know that many plant populations may be assemblages of locally inbred neighborhoods (e.g., Schaal, 1980; but see Lanner, 1966; Levin, 1981) that may meet conditions favoring female production when reproductive effort is great

and male production when the effort is small (Werren, 1980; Fischer, 1981). In any case, an individual's optimal production may depend on the sex ratio produced by other members of the population (MacArthur, 1965; Verner, 1965; Herbers, 1979; Werren, 1980). Sex ratios can also vary with social behavior and related phenomena (Hamilton, 1972; Alexander, 1974; Leigh *et al.*, 1976; Trivers and Hare, 1976; Alexander and Sherman, 1977; Bulmer and Taylor, 1980). In spatially or temporally variable environments conditions may differentially affect the fitness of individuals of different sexes, favoring sex-biased production (Werren and Charnov, 1978; Charnov *et al.*, 1981; Bull, 1981). It is now abundantly clear that sex ratios can be controlled in ways that can lead to unequal expenditures for any given set of parents but that may be adaptive nonetheless (Trivers and Willard, 1973; Howe, 1976; Bull,1980; Nichols and Chabreck, 1980; Burley, 1981a, 1982). Thus there are many reasons for expecting deviations from Fisherian conditions in dioecious species, and similar reasoning could be applied to hermaphroditic species (Charlesworth and Charlesworth, 1981; Fischer, 1981). Enormous variations in effective floral sex ratios are known for a number of species (Schemske, 1978; Lemen, 1980; Primack and Lloyd, 1980; Waller, 1980; Freeman *et al.*, 1981; Willson and Ruppel MS).

When flowers are hermaphroditic, an increase in pollen production may be linked developmentally with an increase in ovule production, and male reproductive effort thus may be more closely tied to resources than is true when sexes are separated. By failing to mature these "extra" ovules and avoiding the major costs involved in seed maturation, however, a plant perhaps could afford to increase male function without a great resource drain. With unisexual flowers on monoecious plants an increase in numbers of male flowers is likely to necessitate a decrease in female flowers, if expenditure on the two sexes draws on the same resources. Male flowers, however, are typically smaller, containing less biomass and fewer calories and sometimes less of other materials as well (Lovett Doust and Harper, 1980). Therefore, in monoecious species it should be possible to increase pollen production with less decrease in ovule production than would be the case if male and female flowers were more similar in cost.

In sum, because gains in fitness from unequal expenditures on the two sexes and highly uneven progeny sex ratios can be great, and because differences in per-flower costs permit a relatively large gain in maleness for a small loss in femaleness, the argument for strict resource-dependence of male RS in hermaphrodites is weakened considerably.

(2) Male RS increases with increasing pollen production, and to the extent that pollen production is resource limited, male RS is also. When pollen deposition is random and the number of available recipients and other factors remain unchanged, however, a two-fold increase of pollen released by an individual results in only a 50% increase in the number of stigmas pollinated. Thus there are rapidly diminishing returns for increasing pollen production. Even if pollen production is governed through resource allocation by the individual and hence eventually by resource availability, it is always true that male RS is limited by the number of matings obtained. In contrast, female RS seldom seems to be limited by the number of pollen grains received and it is likely to increase with increased resources more rapidly than does male RS.

CHAPTER THREE

Male-Male Competition and Female Choice: Bases and Mechanisms

In the first part of this chapter we discuss the bases for female choice: the general utility of genetic quality and male parental investment and the use of pre- and postzygotic mechanisms. We then detail some possible means by which males might influence female choice postzygotically, at the cellular level. This establishes potential mechanisms of influence on zygote fate by males as well as females. There follows a discussion of several events at fertilization or during early development that may provide additional mechanisms of zygote or embryo control and for which sexual selection may provide an evolutionary explanation.

3.1 GENERAL ASPECTS OF FEMALE CHOICE

Female plants, and the female functions of hermaphroditic plants, are likely to evaluate certain aspects of mate quality, including the genetic quality and complementarity of male gametes from various sources. At its most rudimentary level, appropriate mate selection involves species recognition. More sophisticated evaluation may include recognition of genetic similarity, indices of vigor (e.g., growth rate of pollen tube: see Russell, 1980), stamina (ability to maintain fertility), and competitive ability in "male-male" (pollen-pollen) encounters. Pollen competition may be physical (e.g., rapid growth to effect fertilization and/or to prevent other males from doing so) and chemical (suppression of activity of other males). Substantial resources for pollen tube growth are derived from the pistil in some angiosperms (LaBarca and Loewus, 1973; Rosen, 1975), although reserves in the pollen grain are used at first (Jones, 1928). Could it be that males can extract female resources to enhance the rate of pollen tube growth? Alternatively, can females control the amount of resources provided for pollen tube growth? As in animals, it may be advantageous for females to prefer competitive males in order to produce sons that will be superior competitors (Fisher, 1958; Trivers, 1972); if so, intersexual and intrasexual components of sexual selection will evolve to reinforce one another (Darwin, 1871; Fisher, 1958). Specific measures of quality will vary from species to species.

Genetic complementarity is probably more difficult to detect prezygotically than are some other measures of quality, and females of many species may therefore practice abortion of zygotes having nonharmonious combinations of genes. Bertin (1982c) found in hand-pollinated *Campsis radicans*, which is hermaphroditic and predominantly self-incompatible, that the success of pollen donation tended to be reciprocal between plants; that is, if a particular plant is a poor donor for another, then it is likely to be a good recipient from that plant. He suggested that lack of fruiting resulted from abortion rather than failure to fertilize (see also Crowe, 1971). This lends support to the idea of genetic complementarity; if it were simply a matter of cross-compatibilities, the detection should occur prezygotically (de Nettancourt, 1977).

A primary aspect of genetic complementarity is the similarity of parental genotypes (see section 2.1). The optimal level of outbreeding should vary within and among species. This criterion seems particularly relevant to plants because of their immobility; active search or migration is impossible, and it is likely that in many species neighbors and potential mates tend to be relatives. Factors that favor different degrees of tolerance to selfing were discussed above. When routine inbreeding is not disadvantageous, theory predicts that selfing or parthenogenesis should evolve to escape at least some of the costs of sexuality, especially the cost of meiosis and that of mate-getting (Williams, 1975; Solbrig, 1976; Bengtsson, 1978; Maynard Smith, 1978). (Parthenogenesis also frees the plant from recombinational load.) That the sexual habit is costly suggests that maintained, habitual breeding between close neighbors and relatives is suboptimal. If so, traits that reduce the fraction of a female's reproductive effort committed to unions with near neighbors will be selected for.

The notion that individuals of any species might select mates using a criterion of genetic quality has been questioned recently on the grounds that the heritability of fitness tends to be low or nonexistent. Most references to empirical measures of the heritability of fitness components (e.g., Williams, 1975; Maynard Smith, 1978; Borgia, 1979; Harpending, 1979) cite Falconer (1960), who collected only a few measurements of traits with probable implications for fitness. Most of these studies involved domesticated animals, among which much additive genetic variance may have been lost (R. Doyle, pers. comm.). More recently, relatively high heritabilities have been discovered for some components of fitness (e.g., Istock, 1978; Giesel and Zettler, 1980; Grant and Price, 1981; Boag and Grant, 1981; Giesel *et al.*, 1982; but see Mukai *et al.*, 1972; and Falconer, 1981). Moreover, it has been demonstrated repeatedly that additive genetic variance for traits, including those with probable fitness consequences, remains higher in experimental populations subject to temporal and/or spatial variability than in populations kept under constant conditions (MacKay, 1981, and references therein). It remains to be empirically determined how high the heritability of fitness must be for individuals to be capable of using genetic "quality" as a basis of choice.

Related to the criticism that the heritability of fitness is too low for gene quality to be "useful" in mate choice is the opinion that sexual selection leads to the rapid fixation of traits, at which time heritability of those traits drops, of course, to zero. It is not clear, however, why sexual selection should be any more effective in eliminating intrapopulational variation than are other aspects of natural selection. Why, for example, would a hypothetical female preference for "speedy" males be any more effective in fixing genes for speediness than predator pressure acting on both sexes? The logical extension of this argument is that, because natural selection should be effective in fixing genes, evolution through natural selection should come rapidly to a halt, to the discouragement of all researchers interested in the dynamics of evolution, not just those who study sexual selection.

We take the possibly unfashionable position that intraspecific competition should tend to accelerate evolution rather than to stop it. That is, both intrasexual and epigamic selection should operate so that the most competitive and/or attractive individuals at any one point in time have the greatest mating advantage. As a result of directional selection, over evolutionary time the phenotype of the most competitive individuals should change. The process is parallel in conception to one that apparently has resulted in increasing brain size among mammalian predators. Predators and their prey are continually selected to "best" one another—prey to avoid being killed, predators to find more efficient prey-capturing techniques (an interspecific "arms race"; see Dawkins and Krebs, 1979). It appears that mammalian predators have been selected to "outwit" their prey, and this has resulted in ever-increasing cranial capacity (Jerison, 1973; Gould, 1977).

The ultimate way to establish whether or not females can

benefit from choosing mates on the basis of genetic quality is to compare the fitness of females allowed no choice of mates versus that of those permitted substantial choice (either male parental contributions must be controlled for, or species studied in which males make no such contributions). In what apparently is the only such experiment yet performed, intraspecific competitive ability of "offspring" (first instar larvae) was found to be greater for female fruit flies that were allowed a choice of mates (Partridge, 1980). The need for additional research along these lines is clearly evident. Additionally, there is appreciable evidence to suggest that organisms make mating decisions on the basis of genetic constitution, although the fitness consequences have not been determined. In one moth species that is polymorphic for a locus thought to display heterosis, Sheppard (1952) found that females homozygous for the locus preferred males of the opposite homozygous genotype, thus ensuring heterozygosity of their offspring. Mate choice for characteristics having a probable or well-established genetic basis has been found in insects (e.g., Merrell, 1949; Bastock, 1956; Maynard Smith, 1956; Parsons, 1973; Cade, 1979, 1981), fish (Semler, 1971), birds (Cooke *et al.*, 1972, 1976; Burley, 1981), and mammals (Yanai and McClearn, 1972, 1973; D'Udine and Partridge, 1981). It seems most unwarranted to assume that such preferences generally lack associated implications for offspring fitness.

If genetic quality and complementarity are important aspects of mate choice, a question may arise: perhaps this is merely an issue of selection for outbreeding and not a component of sexual selection. Were outbreeding of selective advantage per se, the fitness gains for male and female should be the same. In the context of sexual selection, however, we would expect the fitness gains of the sexes to differ and, further, would expect to find evidence that males conflict with females over acceptability of zygotes. Thus we predict that males will tend to counter female tactics and that the pollen most able to germinate, even on the stigma of genetic relatives, will be favored by selection.

In addition to genetic quality and complementarity, a possible third criterion of mate choice exists for animals and also for plants: selection on the basis of material resources a mate is willing to contribute to an offspring's growth or probability of survival (Trivers, 1972). Mating and paternal investment by male animals are well known; such investment in plants has not been considered previously but is discussed below. Here we have defined male investment in plant zygotes as mating investment; it provides a criterion for choice by females and functions in male-male competition. The benefits to females of aborting relatively well developed embryos are diminished if their future energetic requirements are comparatively small (Trivers, 1972; Dawkins and Carlisle, 1976; Boucher, 1977); that is, offspring requiring less future investment (as a result of male investment) are more valuable than those requiring more. Given a large nonadditive component to the heritability of fitness, it can be argued that investment is more predictable than genetic quality of offspring and that females should therefore tend to weigh male expenditure heavily (Maynard Smith, 1978). Maynard Smith (1978) argued that the large nonadditive component also favors mating with unlike individuals; if this is true, then the relative importance of investment versus genetic traits may increase as opportunities for outcrossing diminish. Pollen grains do contain metabolic machinery, stored nutrients, and other material that can enter the egg at fertilization (see below) and might contribute to zygote growth. Furthermore, some pollens contain hormones that prevent abscission of the flower and stimulate growth of the fruit (Nitsch, 1971), but we do not know if the pollen of different individuals differs in the dosage or activity of such hormones.

Considerations that affect the amount of investment males may adaptively place in offspring include trade-offs between content or size and number of pollen grains (the more resources per pollen grain, the fewer can be made for the same total allocation) and certainty of paternity (it is maladaptive to invest energy in the offspring of other males: see Trivers, 1972; Alexander and Borgia, 1979; Warner, 1980). In wind-pollinated higher plants an additional limit to male gametophyte size is probably set by the diminished dispersal range of large particles (e.g., Whitehead, 1969; Gregory, 1973; Levin and Kerster, 1974). The type of material packaged as an energy source affects pollen size and possibly pollen density. "Starchless" pollen, common in entomophilous species, contains oils and sugars (perhaps as a reward for pollinators) and tends to be smaller than "starchy" pollen usually found in anemophilous species (Baker and Baker, 1979). Although materials contained in pollen grains may function in pollinator rewards or in maintenance of the pollen grain itself, they may also be put into the egg at fertilization and perhaps used for zygote growth (see section 3.2). Trade-offs between production of large, possibly less dense pollen versus smaller, starchless pollen are therefore likely to be complex.

Although pollen grain diameter is often used as a taxonomic character, intraspecific and intra-individual variation in pollen size is sometimes considerable (Müntzing, 1928; Wodehouse, 1935; Cain and Cain, 1944, 1948; Jones and Newell, 1948; Bell, 1959; Faegri and Iverson, 1964; Davis, 1966; Bassett and Crompton, 1968; Figure 1). Inbreeding decreases pollen size in maize (*Zea mays,* see Johnson *et al.*, 1976). Temperature, water availability, and nutrition may substantially affect pollen size (Schoch-Bodmer, 1940; Mikkelsen, 1949; Wagenitz, 1955; Bell, 1959). Geographic clines in pollen size occur (e.g., Cain and Cain 1948), and pollen collected at the same site may vary in size from year to year (e.g., Jones and Newell, 1948). These findings suggest that males may not benefit from minimizing pollen size; otherwise, variability should be quite small,

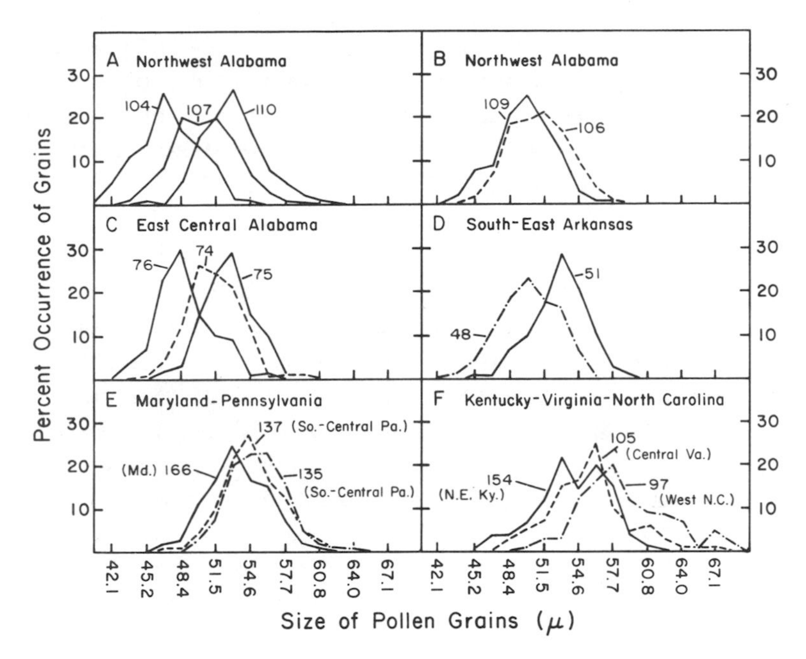

FIGURE 1. Variation in size of pollen in *Pinus echinata*. Data from Cain and Cain (1948). Each numbered distribution represents the diameters of 150 grains from one pollen collection. Separate collections at nearby sites are illustrated in *A* and *C*. Separate collections at a single site are illustrated in *B, D,* and *E* (Pennsylvania data only).

at least within individuals. This opens the possibility that males are selected to endow their pollen with substances that enhance the likelihood of acceptance. When females are resource limited, the collective effects of paternal contributions from numerous pollen grains may increase significantly female RS. Pollen endowments could also increase competitive ability in pollen-pollen encounters and thus influence female choice in a second way.

Mechanisms of female choice of mates can be broadly classified as prezygotic and postzygotic. One prezygotic mechanism employed by some hermaphroditic plants involves morphological and temporal arrangement of male and female function to affect the incidence of self pollen arriving at stigmatic surfaces. Furthermore, prezygotic recognition of pollen types has been demonstrated for many angiosperms (Lewis, 1949, 1954; Bateman, 1952a; Arasu, 1968; Heslop-Harrison, 1975; Lundqvist, 1975; de Nettancourt, 1977); females possessing the ability to assess pollen quality may act subsequently to inhibit or enhance fertilization by particular pollen. Pollen interactions with the style in *Zea mays*, for instance, appear to favor pollen tubes of genotypes that result in greater seed and seedling size (Ottaviano *et al.*, 1980). Prezygotic decisions are sometimes made difficult by interactions among pollen types, however. If pollen grains of differing compatibilities with the stigma are growing simultaneously, compatible types may improve the acceptance of incompatible types (e.g., Gilissen and Linskens, 1975; Pandey, 1979). Such effects are often dosage dependent and can be seen even in interspecific combinations. These observations were made on unnatural crosses and may not reflect evolved responses. They raise the possibility, however, that males can interfere chemically with each other's success, as well as "deceive" females about their identity. Dosage dependence of the effects might make it advantageous for pollen to be deposited in clumps (and see Willson, 1979). In some cases, however, the converse is true: incompatible pollen grains may interfere with compatible grains (Ockenden and Currah, 1977).

The "pollen population effect" in angiosperms is well known and widespread (Brewbaker and Majumder, 1961; Brewbaker and Kwack, 1963; Jennings and Topham, 1971; Cohen, 1975) but perhaps not universal (Levin, 1975). Species or even conspecific populations commonly differ in the minimum number of pollen grains required to effect fertilization, and the shape of the response curve relating pollen abundance to seed set also differs (Brewbaker and Kwack, 1963). It would be fascinating to discover if sibling pollen grains arriving together on a stigma ever exhibit a greater synergistic interaction than nonsibs (see also Kress, 1981). Another consequence of varying pollen loads in certain dioecious species is a skewing of the progeny sex ratio: large numbers of pollen grains produce an increased proportion of females (Correns, 1921a, 1921b, 1922, in Jones, 1928), with unknown consequences for the fitness of the pollen donor(s).

The prezygotic detection abilities of gymnosperms are much more limited. *Inter*specific prezygotic incompatibility has been found in the genus *Pinus* (Hagman, 1975) and in *Picea* (Kossuth and Fechner, 1973), but intragenus incompatibilities have been found lacking elsewhere (Hagman, 1975). General lack of prezygotic recognition (including intraspecific recognition) has been reported to be a general gymnosperm trait (Stockwell, 1939; Sarvas, 1962, 1968; Koski, 1971, 1973).

Gymnosperms (Orr-Ewing, 1957; Hagman and Mikkola, 1963; Fowler, 1965b; Lahde and Pahkala, 1974; Stern and Roche, 1974) and many angiosperms (but not all: see Stephenson, 1981) practice zygote abortion, however. Abortion of zygotes of varying quality is a potentially important postzygotic mechanism of mate selection. In order to utilize abortion for this purpose, females must produce a greater number of ovules or archegonia than they can expect to mature (and see Lloyd, 1980a).

At least two considerations should affect the fraction of zygotes to be aborted: the amount of resources available and the attributes of pollen received. If these are predictable prior to ovule formation, plants may be selected to abort a constant fraction of their zygotes. Estimations of resource availability should improve, sometimes, as a season progresses, and females may respond to improved information by increasing or decreasing the fraction to be aborted. Sequential maturation of ovules (Mikkola, 1969) may also constitute a response to the problem of making accurate assessment of resources early in the breeding season. Preferential abortion of young fruits, due to competition with older ones, is well known in some species (Addicott and Lynch, 1966; Kozlowski and Keller, 1966; Herbert, 1979; Tamas *et al.*, 1979; Lee, 1980). *Cassia fasciculata* preferentially aborts pods with lower percentages (not merely numbers) of fertilized ovules (Lee, 1980). Such assessments may occur also at the ovule level.

Even if females possessed accurate information about resources early in the season, they may nevertheless be selected to vary abortion rates in response to the quantity and quality of pollen received. We define a good pollen year to be one in which high quality pollen is abundant. If such years are not frequent, maximal reproduction should be favored when they occur. By contrast, during poor pollen years, it may often be advantageous to restrict reproductive effort to the extent that resources can be stored for future use. Therefore, if either resources or pollen is somewhat unpredictable, it should be advantageous even for females with sophisticated prezygotic selection abilities to abort ovules after fertilization and to display flexibility in abortion rates.

Regardless of whether females practice pre- or postzygotic selection, or both, and whether they abort a fixed or variable fraction of zygotes, their evaluation of the acceptability of pollen of a given quality should depend upon availability of pollen of higher quality. This conclusion is parallel to one reached in several models of optimal foraging for prey types (MacArthur and Pianka, 1969; Schoener, 1971; Pulliam, 1974; Charnov, 1976). For simplicity, let us assume that resource levels remain constant. In the poorest pollen years females must accept low-quality pollen or opt not to reproduce at all. During intermediate pollen years only limited amounts of high-quality pollen are available, so females will accept a lower proportion of poorer-quality pollen. In very good pollen years the highest quality mates (and/or zygotes) are available and females need not make offspring from inferior pollen (Figure 2). If resources and pollen were to become abundant simultaneously, females might mature a large seed crop including zygotes, fathered by low-quality pollen, that would otherwise be aborted.

Thus the acceptability of pollen of any given quality should vary with both pollen and resource availability. Degrees of outcrossing and the acceptability of inbred pollen are known to vary from year to year and population to population (Dilman, 1938; Claassen, 1950; Bateman, 1952b; Gutierrez and Sprague, 1959; Harding and Tucker, 1964; and see references above, section 2.1). Most studies do not adequately control the availability of either pollen or resources, however, which makes sound interpretation impossible.

The decision to reduce seed crop size or to defer repro-

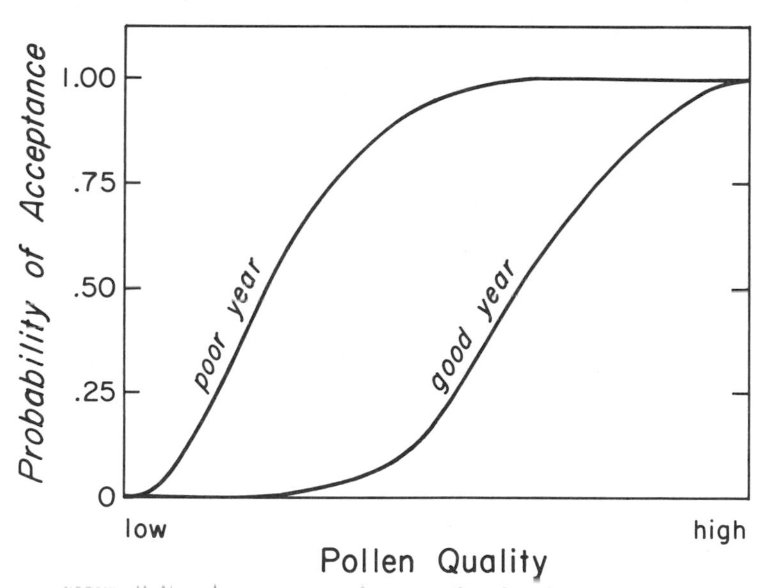

FIGURE 2. Female acceptance of mates. When females choose mates based on the availability of those of higher quality, the shape of resulting pollen acceptance curves varies with the composition of pollen types present. If the highest quality present is abundant, lower-quality types are much less likely to be accepted than when the best types are less available.

duction in poor pollen years should depend upon the probability of a female's survival to the next breeding season, upon her ability to store resources for future use, and the probabilities that pollen resources will improve and that other resources will be abundant in the following breeding season. Probability of survival may only rarely approach certainty, and females are likely to be less than totally efficient at storing resources for long intervals. Also, their information regarding the subsequent year's resources must be very limited. In many instances, then, females will not forego reproduction entirely if they have substantial resources available. As a result, they may often be forced to accept a number of suboptimal mates.

Implicit in the hypothesis that optimal mating decisions are based on levels of receipt of superior pollen is the necessity that individuals possess the ability to evaluate and integrate "mate" (pollen and/or zygote) quality over some fraction of the plant. Integration at a plant-wide scale would sometimes be most efficient, for the distribution of pollen quality is likely to be patchy. Reproductive resources such as starch and nitrogen are known to be diverted to fruits from all parts of the plant (e.g., Gäumann 1935; Kramer and Kozlowski, 1960; Leonard, 1962; Mochizuki, 1962; Dickmann and Kozlowski, 1968; Mooney and Hays, 1973; Lewis *et al.*, 1964; Rockwood, 1974). In fact, carbohydrates are sometimes exchanged between plants through root grafts (Bormann and Graham, 1959; Graham and Bormann, 1966; Bormann, 1961, 1962, 1966; Kozlowski and Cooley, 1961; Lanner, 1961; Saunier and Wagle, 1965; Wold and Lanner, 1965). Therefore, plant-wide integration of mate choice is not out of the question.

Several factors may limit the possibility or utility of plant-wide integration, causing the decision-making process to be controlled at a more local level. Since the mode of integration is presumably chemical, body size must strongly affect the possibility and speed of communication: the larger the plant, the more time consuming the evaluation process.

In turn, limits to the duration of the selection process may be set by environmental constraints favoring rapid zygote maturation. Additional time constraints may be set by males if, as we hypothesize below, males compete to enhance their probability of fertilizing females or if they "attempt" to prevent abortions. Localization of resources may also limit the utility of plant-wide selectivity. Resources such as light (Kozlowski and Keller, 1966; Smith, 1981) may not be uniformly available over a plant's surface, and concurrent photosynthesis is sometimes the carbon source for seed production (references in Willson, 1972). Hormones released by embryos may affect resource allocation, but only at a local level (e.g., McCollum, 1934; Dearborn, 1936; Nitsch, 1971). Certain reproductive organs may be better supplied with growth hormones than are others (e.g., tomatoes: see Wittwer, 1943). If resources cannot be efficiently transported from one part of a plant to another, then allocation to ovules must be made at a local level. Resource availability should affect the incidence of abortion, so selection of zygotes must be practiced on a scale at which resources can be transported efficiently.

The mate selection process occurs in stages, and it is likely that different stages operate at different scales. Early decisions may be made at a very local level, in part because of limited availability of information. For example, prezygotic detection and inactivation of self pollen may allow a typically outcrossed species to minimize the incidence of inbred zygotes without requiring information on quality of pollen elsewhere. Decisions occurring after fertilization may involve broader scale comparisons regarding the relative quality of offspring of alternative mates.

3.2 CELLULAR MECHANISMS OF ZYGOTE CONTROL

We have argued that females may use differential abortion as a mechanism for choice among the potential fathers for her brood. In the face of possible elimination by the female, males that manage to improve the survival of their zygotes would be selected for. This section explores mechanisms by which males might conflict with females for control of zygote fate.

Theory leads us to expect *maternal* control of zygotes because females in outcrossing species typically invest more in offspring than do males. In fact, many instances of female control are known. The egg contains maternal cytoplasm and organelles, which frequently are incorporated into the zygote. Maternal gene products commonly operate during early embryogeny (e.g., Davidson, 1976; Grant, 1978; Davenport, 1979). In some cases paternal chromosomes are even eliminated or inactivated, as in *Ascaris*, some insects and protozoans, and the X chromosome of certain mammals (Sharman, 1973; White, 1973; Gurdon, 1974; Lyon, 1974; Canning and Morgan, 1975; John and Lewis, 1975; D'Amato, 1977; West *et al.*, 1977). Some evidence, however, indicates that females do not make all decisions regarding male contributions to zygotes. Although the physical exclusion of male organelles by the egg is accepted as a common occurrence, in certain angiosperms the organelles are debilitated even before fertilization (in fact, during microsporogenesis) so that the male is in proximate control of their fate (Vaughn *et al.*, 1980). At least in these species, then, the male may make no "attempt" to invest extrachromosomal materials in his zygotes. Furthermore, there may be selection for smaller, more streamlined sperm in species that inject sperm via pollen tubes, as a means of more efficient injection (Kirk and Tilney-Bassett, 1978: 428; and see below); this might prevent the evolution of high levels of male investment.

Nevertheless, it seems possible that a male may have several potential means of influencing his zygotes in ways that could influence the controlling female's choice in his favor. If males could speed up the development rate of their offspring, either through genetic means or by investing materials in their zygotes, it would become unprofitable for a female to abort these well-developed embryos because they would require less additional investment by the female than more retarded embryos. Furthermore, a genetically inferior male may enhance his quality by providing resources that reduce female investment in his zygotes. (It is also conceivable that male investment in zygotes somehow wrests the ultimate control from the female. For reasons just outlined, we suspect that this possibility is less likely than males influencing female choice.)

To examine the potential of such means of controlling zygote fate via male investment patterns, we need to establish at least three things: (1) possible mechanisms by which influence could be exerted; (2) the existence of variation among individuals in degree of expression of such mechanisms (this establishes the array from which females may choose and the basis for future natural selection; lack of variation, however, might mean only that all the selection occurred in times past); and (3) the actual effect of such mechanisms on zygote survival. We have examined selected literature in genetics and development in search of material relevant to these three points; in no single case is there evidence for all three. Nevertheless, there is some evidence from different organisms for at least the first two of these. We first present a review of certain nuclear events. Because these are mostly from the zoological literature and there is little concrete evidence of similar events in plants, the function of the review must be largely suggestive. Next, we review a variety of cytoplasmic phenomena; here there is considerably more botanical literature.

Nuclear phenomena

First we will examine features relevant to the regular chromosome complement of the organism—that is, the "*A* chromosomes" that generally behave as expected at mitosis and meiosis, that carry most of the active genes, and that are the typical subject of study for most geneticists. Both interspecific and intraspecific crosses and nuclear transplants offer potentially useful information; and studies of morphogenesis and of allelic activation are also germane. Changes in rates of development are more likely to be relevant to our concerns than changes in morphogenetic patterns, because rapid development of a zygote seems to be a possible way for the zygote to win more maternal resources than the other zygotes and provide a basis for female selectivity.

In many animals very early development depends on material synthesized by the mother and deposited in the egg during oogenesis. Although new RNA is commonly synthesized by the early zygote, it may not be used until later (except in mammals, which have tiny eggs) (e.g., Davidson, 1976; Davenport, 1979; Magnuson and Epstein, 1981). When the zygote genome (in an interspecific hybrid) turns on, in most cases maternal and paternal alleles are activated synchronously, although occasionally the maternal ones precede the paternal (Champion and Whitt, 1976, and references therein). To date, there is no evidence that the paternal genome ever switches on before the maternal genome in the zygote. Rather few organisms and alleles have yet been studied, however, and the possibility cannot be excluded.

Developmental rates might be influenced by paternal input. If an egg is fertilized by foreign rather than conspecific sperm in an experiment using interspecific hybridization, changes in rates of cleavage may be observed, as in teleosts (Newman, 1914). But Newman noted that such effects may not be hereditary; a change did not necessarily reflect the male's genome, merely his foreignness.

In both forced hybridization and nuclear transplant experiments between species, the greater the taxonomic distance between the parents, the earlier male genetic effects can be detected. This may merely reflect the degree of difference in developmental patterns rather than an actual difference in time of genome activation. It is conceivable, however, that the decline of female ability to suppress male influence reflects their shared evolutionary history: mechanisms of suppression by females are more likely to be effective among closely related species. Typically, these experiments demonstrate taxonomic differences in apparent time of activation of the zygote genome but do not help much in discriminating male from female genomic effects; it is essential to know which nonpaternal effects are due to the zygotic maternal genome and which to maternal influences established during oogenesis.

Nuclear transplant experiments placing a haploid nucleus of one species in the enucleated egg cytoplasm of another demonstrate that both male and female haploid genomes are capable of regulating normal development to some point, which differs among taxa (e.g., Fankhauser, 1955). This kind of experiment does not necessarily tell us what the normal relationship is, however, when both parts of zygotic genome are present and both are of the same species as the cytoplasm. For all interspecific crosses and transplants, the possibility of "trauma" due to disruption of nucleocytoplasmic coadaptation may obscure normal events (Gurdon and Woodland, 1968; Lewin, 1974; Bell, 1975; Grun, 1976). Nevertheless, evidence from these experiments may at least be useful in suggesting the existence of intertaxon differences in time of onset of zygote genome activity (see also Grant, 1975: 441) and in the capacity of each haploid genome. Such differences might reflect different evolutionary balances in male-female interactions.

Intraspecific studies of allelic isozyme activation, like the interspecific studies, generally indicate that maternal and paternal genomes turn on synchronously in the zygote; in one case the maternal one precedes (Ohno *et al.*, 1968; Klose and Wolf, 1970; Engel and Wolf, 1971; Wolf and

Engel, 1972; Engel, 1973; Champion and Whitt, 1976). Thus, as before, there is no evidence of paternal genetic precedence in the zygote, but again the sample is still very small.

Intraspecific crosses between strains have been made in both mice (laboratory *Mus musculus*) and domestic rabbits (*Oryctolagus cuniculus*); these strains differ in body size and/or rate of zygote cleavage. Mammals are unusual among animals in that net growth of the embryo begins very early; RNA synthesis and use by the zygote also occur in the early stages of development. For the rabbit, the studies hint that there is a male effect on rate of cleavage in the hybrid zygote, but the effect is small and so is the sample (Castle and Gregory, 1929; Gregory and Castle, 1931). Later work by Castle (1941) shows that the female has more effect on zygote cleavage than the male and that much of the observed effect is nonnuclear. For our purposes we must, therefore, dismiss this rabbit work as mostly wishful thinking.

Conflicting reports are found for mice. In one study (Whitten and Dagg, 1961) different strains had different cleavage rates. Reciprocal hybrid crosses revealed that the fast strain was expressed more in the zygote, regardless of which parent was of that strain. A second study (Bowman and McLaren, 1970) used intrastrain lineages selected for large and small body size but failed to find accompanying differences in developmental rate. Furthermore, there was considerable variability among females of each lineage, suggesting influence of other factors. Thus neither of these studies provides evidence for specifically paternal effects on cleavage rates. Manes (1975) asserted that the earliest expression of the male genome of the laboratory mouse is in the morula stage.

A number of organisms (for animals, see White, 1973; mosses, see Vaarama, 1953; higher plants, see Battaglia, 1964, and see below) have been shown to have, in addition to the *A* chromosomes, a number of accessory or *B* chromosomes. Genes in *B* chromosomes are not usually very active, but the chromosomes can affect the frequency of chiasma formation and rates of mitosis (Barlow and Vosa, 1970; Rees, 1972; Vosa, 1972; Vosa and Barlow, 1972; R. N. Jones, 1975, 1976). Nongenetic nuclear phenomena, such as the long-known relationships among DNA content, nuclear volume, and cell size (Price *et al.*, 1973), may have enormous evolutionary consequences (e.g., Bennett, 1972; Hinegardner, 1976; Cavalier-Smith, 1978; Willson, 1981). Such effects may often reflect changes in ploidy levels of the *A* chromosomes but perhaps can be modified by *B* chromosomes in the nucleus. Heritable changes in DNA content can even be induced environmentally (Durrant, 1962; Evans, 1968; Durrant and Jones, 1971).

Unlike the *A* chromosomes, the number of *B* chromosomes can vary greatly both among populations and among individuals (Fröst, 1957; Jones, 1975). Such differences are sometimes related to ecological or geographic variables (e.g., Sparrow *et al.*, 1952; Müntzing, 1954, 1957, 1966; Fröst, 1957; Hewitt and Brown, 1970; Moir and Fox, 1977). Furthermore, at least in some angiosperms *B* chromosomes are differentially transmitted during meiosis to the cells that become functional gametes, either male or female, but usually not both (Müntzing, 1954, 1966; Bosemark, 1956a,b, 1957; Fröst, 1959; Kayano, 1962). In *Xanthisma texanum B* chromosomes are present in the early embryo but are eliminated in the late embryo (D'Amato, 1977), suggesting that their time of utility is then past. Fragmentary chromosomes in *Trillium grandiflorum* segregate differentially into the future endosperm (Rutishauser, 1960). *B* chromosomes sometimes slow rates of mitosis out of proportion to their effect on cell and nuclear sizes (Ayonoadu and Rees, 1968; Evans *et al.*, 1972). Because they are often heterochromatic, they also may have effects on cell volume without affecting mitotic rates (Nagl, 1974; Nagl and Ehrendorfer, 1975). *B* chromosomes in rye (*Secale cereale*) affect rates of pollen tube growth, though the magnitude of the effect is not directly related to dosage of *B*'s (Puertas and Carmona 1976). Although the effects of *B* chromosomes on the development and fate of zygotes seem not to have been investigated, their effects on cell division, their variation among individuals, and their patterns of transmission during gametogenesis all suggest that they could have important effects on the zygote.

Gene amplification through doubling of chromosome numbers by a haploid sperm before it fertilizes an egg may increase template availability for synthesis of RNA (D'Amato, 1977). This phenomenon is known or suggested for a variety of animals, a fern, a liverwort, and some grasses (references in D'Amato, 1977). Chromosome doubling is also reported from certain animal eggs and thus may occur on the female side as well. That these doublings could have evolved through male-female interactions seems very possible.

In sum, neither the gene activation literature nor that on nuclear functions in development has yet provided good evidence for possible male-male or male-female interactions of a sociobiological sort, although the potential exists. The best nuclear evidence so far seems to be provided by the literature on gene amplification in gametes and on *B* chromosomes.

Cytoplasmic phenomena

For the envisioned cytoplasmic events to transpire, male cytoplasm and its included organelles must enter the egg. This happens in many animals and is very common in plants: the lower archegoniate plants, gymnosperms, and some angiosperms (Appendix; Ferguson, 1913; Wilson, 1925; East, 1934; Vazart, 1958; Steffen, 1963; Austin, 1968; Grant, 1975; Pfahler, 1975; Beale and Knowles, 1978; Eberhard, 1980; Russell, 1980). Because male cytoplasmic contributions to eggs may suffer various fates (Birky, 1976), including destruction at one extreme and multiplication at the other (e.g., Favre-Duchartre, 1966, 1974), mere entry of male material is insufficient information for final interpretation. Even the products of destruction might be recycled usefully, however.

If male cytoplasm and its inclusions enter the egg at fertilization and persist, what possible effects could be imagined? There are at least four possibilities relevant to our model: male material could (1) block female RNA or proteins, (2) synthesize products for growth and division of zygote cells, (3) bring in heritable factors, and (4) induce the female to synthesize materials for zygote growth. Let us take the first three in order (we can find no information regarding the fourth).

(1) Because much zygote RNA is synthesized early in development but is then tied up and stored for later use (Grant 1978), clearly there is a mechanism for blocking RNA. Furthermore, in at least one genus (*Gossypium*), zygotes routinely tie up the original maternal ribosomes before the first cell division and zygote RNA takes over (Jensen, 1974). We have not been able to find any evidence specifically for male blockage of female RNA, but this mechanism cannot be excluded.

(2) Paternal organelles and molecules could contribute to synthesis of metabolic products in the zygote. Nonchromosomal DNA multiplies during oogenesis and makes RNA (John and Lewis, 1975); perhaps similar events could occur during male gametogenesis and male gametes could donate such DNA and/or RNA as machinery for synthesis in the zygote. Ribosomal RNA synthesized by nurse cells and transferred to the oocyte in certain insects and mollusks and by follicle cells in some mammals may be a substitute for gene amplification and resultant synthesis (Raven, 1961; Gall, 1969; Gurdon, 1974; Cohen, 1977). Thus it seems to be possible for one cell to supplement its own synthetic

machinery with that of another. Levels of RNA (at least some types) may be indeed important: experimental injections of messenger RNA into cells can increase synthesis of products (up to a point) without interfering with synthesis of mRNA that was already present. In short, the rate of protein synthesis is, in this case, limited by the availability of mRNA (Gurdon, 1974). The embryonic RNA in *Cephalotaxus drupacea* may be largely of male origin (Gianordoli, 1974). Such male-derived activity could even influence nucleocytoplasmic interactions, affecting the maternal or zygotic nuclear genome as well. Furthermore, different genotypes of certain species have different numbers of genes coding for ribosomal RNA (R. N. Jones, 1976), suggesting a basis for individual variation.

In ferns paternal microtubules enter the egg and are said to be important in growth and development (Bell, 1979); other materials also enter the egg of at least two species (Duckett and Bell, 1971; Bell and Duckett, 1976; Myles, 1978). Male mitochondria of several gymnosperms (and perhaps some angiosperms—Anderson 1936; Linskens 1969; Pfahler 1975) enter the egg with male cytoplasm (Favre-Duchartre 1957; Camefort 1968, 1969; Chesnoy 1969, 1973; Chesnoy and Thomas 1971; Gianordoli 1974). Although sperm mitochondria are customarily associated with flagellar activity of ferns (Duckett, 1975), they may also enter the egg (Tourte, 1971). Paternal plastids also enter the zygote in some ferns and in many algae (e.g., Yuasa, 1952) and in several gymnosperms (Favre-Ducharte, 1957; Camefort, 1968, 1969; Chesnoy, 1969, 1973, 1977; Chesnoy and Thomas, 1971; Ohba *et al.*, 1971; Gianordoli, 1974). In certain fascinating cases maternal cytoplasm is partially or even totally destroyed. Maternal organelles may survive and enter the zygote (Camefort, 1966b, 1967, 1968, 1969; Thomas and Chesnoy, 1969; Chesnoy and Thomas, 1971); but in *Biota (Thuja) orientalis* all the female cytoplasm and inclusions are destroyed and the embryo gets its cytoplasm and organelles from the male alone (Chesnoy, 1977). Thus in a number of cases, metabolic machinery is a male contribution to the zygote.

Male gametes can also bring in starch grains at fertilization (Coker, 1907; Chesnoy, 1973). The pollen grains of at least some gymnosperms have compartments for storage of starch. Christiansen (1973) proposed that the large compartments of *Pinus* and *Cryptomeria* may be associated with the long delay between pollination and fertilization, as compared with the smaller compartments of *Picea*, which has a much shorter delay (Appendix). Starch in angiosperm pollen may fuel the growth of pollen tubes in long styles (Baker and Baker, 1979). Starch thus may be stored nutrition for the gametes and their companion cells (see also Christiansen, 1972a), perhaps constituting mating investment. When starch grains enter the egg at fertilization, however, they may also contribute to the growth of the zygote. If the donation of starch grains is a regular feature of any species, it probably constitutes mating investment in young. Entrance of male cytoplasm and fat droplets into the egg at fertilization in orchids (Poddubmayer-Arnoldi, 1960) might have a nutritional function. Any food reserves in pollen grains (starch, lipids, sugars: see Baker and Baker, 1979) have the potential for entering the egg and contributing to the growth of the zygote. Polyspermy—the penetration of an egg by several sperm—could perhaps also contribute nutrients and/or machinery. This phenomenon occurs widely in animals, some angiosperms (Austin, 1965), certain gymnosperms (Appendix), and some lower plants (Rickett, 1923; Showalter, 1926, 1927a,b; Atkinson, 1938; Chatterjee and Mohan Ram, 1968) and may be relevant here, especially if all the sperm in an egg come from one male.

If male nutritional products and synthetic machinery contribute significantly to zygote growth, this could provide a basis of discrimination by females among the zygotes of different fathers; females could use relative nutritional contribution as a basis for differential abortion of zygotes. Of course, if females were able to resorb efficiently substances from unsuccessful pollen, the females would, in any case, be able to acquire male products and, therefore, they should profit less from evaluating potential mates on the basis of their contributions. MI could nevertheless function to increase the ability of a male's zygote to prevail over others. Information on intraspecific variability in male cytoplasmic donations and their actual effect on zygote growth and fate would be most useful to evaluate the potential for occurrence of male MI in plants.

(3) As we have seen, cytoplasmic inheritance, in the form of both mitochondria and plastids, varies enormously among taxa. Although *biparental* cytoplasmic inheritance is well known, from the algae to the angiosperms, the frequency varies and in unusual cases male transmission prevails. Eberhard (1980) has discussed the possibility of organelle competition in eukaryotes. Detailed studies of cytoplasmic heredity are of interest here, particularly because there is much evidence for genotypic or strain variability in the balance between maternal and paternal transmissions. Much of the work has been done with microorganisms, which are not directly germane to our concerns but which are nevertheless of interest because they exhibit such variability. We review first the cases for two well-studied microorganisms and then the evidence for higher plants.

Crosses between different pairs of strains of the yeast, *Saccharomyces cerevisiae*, demonstrate that the frequency of uniparental inheritance of mitochondria differs among crosses (Birky, 1975a; Birky *et al.*, 1978). Furthermore, there exists great heterogeneity among the zygotes produced in a single cross or mating, apparently at least partly because organelle genomes are differentially replicated during the vegetative cell cycle, producing variability in the progenies of different cells that initially had the same genotype (Birky,

1975b; Birky *et al.*, 1978). In addition, mitochondria from the two parents do not mix evenly in the zygote, so when it buds off new vegetative cells, the offspring receive different assortments of mitochondria (Birky, 1975b; Birky *et al.*, 1978). The control of such intracross variability appears to be unknown but conceivably could vary with parentage. *Saccharomyces cerevisiae* exhibits interstrain differences in mitochondrial DNA, but the mtDNA of one strain is fully functional in the nuclear background of other strains (Sanders *et al.*, 1976), despite Birky's (1975b) remark that laboratory stocks of *S. cerevisiae* are really hybrids between that and closely related species. Therefore, crosses between strains of this species are not plagued by the problem of nucleocytoplasmic compatibility discussed by Grun (1976).

The green alga *Chlamydomonas reinhardti* likewise exhibits individual variability in the frequency of uniparental transmission, both between crosses (Gillham, 1969) and within the progeny of every kind of cross that has been studied (Gillham, 1978). It is difficult to assign to isogamous organisms the epithets "male" or "female"; categorization by "sex" may often be arbitrary (Birky, 1976), or perhaps made on the basis of relative motility (L. Hoffman, pers. comm.). Even with this difficulty, it is interesting to note that while most cytoplasmic inheritance is from one parent ("maternal") and some is biparental a small fraction comes from the other parent only ("paternal") (Birky, 1978). Similar degrees of variability are reported for two other algae (Birky, 1978). Some of this variability may be a function of age or rearing conditions (e.g., Van Winkle-Swift, 1977), but this aspect has apparently been little studied.

In the angiosperms transmission of male plastids is known in several genera (Hagemann, 1976; Gillham, 1978; Kirk and Tilney-Bassett, 1978) and the possibility cannot be excluded in many other cases (Kirk and Tilney-Bassett, 1978). As is true for *B* chromosomes, plastids may be differentially assorted into the functional gametes (Lombardo and Gerola, 1968; Hagemann, 1976). Perhaps the two best studied genera are *Pelargonium* and *Oenothera.*

In *Oenothera* the plastids of different strains and species have different competitive abilities and appear differentially in hybrid offspring (Preer, 1971; Kirk and Tilney-Bassett, 1978). The hybrid nucleus exerts some control over the relative success of maternal and paternal plastids, suggesting that the suitability of the nuclear background may be important (Kirk and Tilney-Bassett, 1978). Furthermore, relative competitive abilities and levels of nuclear control might reflect the extent to which the partners have had opportunities to evolve responses to each other.

In *Pelargonium* different crosses yield different proportions of biparental, maternal, and paternal inheritance (Tilney-Bassett, 1975; Kirk and Tilney-Bassett, 1978). The hybrid nucleus exerts some control of the balance of transmission frequencies (Kirk and Tilney-Bassett, 1978), but even more striking is the effect, before meiosis occurs (Tilney-Bassett, 1976), of the nuclear genes of the maternal parent. The level of male transmission is controlled primarily by the maternal nuclear genotype (Kirk and Tilney-Bassett, 1978); either male or female transmission may predominate, in contrast to *Oenothera* where maternal control is associated with predominantly female transmission (Tilney-Bassett, 1975). The taxonomic form in question is *Pelargonium zonale*, which is an intergeneric hybrid with many cultivars (Kirk and Tilney-Bassett, 1978), so it is perhaps possible that some of these effects may result from disruption of nucleocytoplasmic coadaptations.

Among the cytoplasmic phenomena, then, several possible mechanisms of male influence exist. There is also variability among individuals and among matings in the degree of male transmission of cytoplasmic organelles. Viewed as units of inheritance and/or bits of metabolic machinery, the potential role of organelles in governing the fate of zygotes seems plausible. The utility of male nutritional products or the possibility of male means of blocking female RNA cannot be excluded.

Thus far we have discussed male contributions to zygotes primarily as possible means of decreasing the probability of abortion by a resource-limited female. It is easy to imagine that the function of investment shifts if the nature of the limitation of female RS changes. That is, where female RS is pollen-limited, a male's contributions might constitute parental investment simply to increase the success of his offspring, rather than to best other males in sexual competition. We would expect male investment to be less, as a rule, when female RS is pollen-limited, because of the probable advantages to males of allocating available resources to production of pollen. Exceptions might occur especially when there is a premium on rapid zygote growth for some ecological reason.

It is pertinent here to review and assess three extant hypotheses regarding the evolution of uniparental versus biparental cytoplasmic inheritance. Sager (1975: 265) proposed that "the evolutionary significance of uniparental inheritance of organelle DNA lies in the maintenance of genetic stability by blocking recombination and perhaps by limiting allelic polymorphism as well." One difficulty with this hypothesis is the implicit dichotomy drawn between cytoplasmic and nuclear inheritance, the latter being typically biparental. The question must then be asked: why should uniparental inheritance (when it occurs) be advantageous for cytoplasmic genes but not for nuclear ones? The observed variability in the frequency of biparental cytoplasmic transmission also places a restriction on Sager's hypothesis, because it suggests that the degree of biparental transmission must be adjusted to a varying "optimum" level of recombination.

Grun (1976: 350) suggested that transmission of cytoplasmic factors through the egg only, in most species, may have evolved as an insurance mechanism to protect the *host* population. Suppose that some cytoplasmic allele had high competitive value and therefore spread through its host. If transmitted via sperm, he argued, it will be passed to many females, and if deleterious represents a "waste in genetic material of the host population." On the other hand, if it occurs in the female only, it can only spread through that lineage. Several ecological and evolutionary difficulties arise with this hypothesis (see also Eberhard, 1980). First, this suggestion takes no account of the known frequency of male transmission—precisely that feature supposed to be so disadvantageous. Furthermore, even if cytoplasmic organelles originated as parasitic microorganisms, there seems little reason to suppose that they still function that way. Also, it is hard to envision natural selection acting to protect any population, much less one different from that under scrutiny (Williams, 1966). To the extent parasites have evolved reduced virulence, it must be because they increase their lifetime reproductive success that way (by not killing off the host too soon). This then points up the fundamental weakness of the hypothesis, namely that it is firmly based in group selection rather than individual selection.

Finally, Tilney-Bassett (in Kirk and Tilney-Bassett, 1978: 428) proposed that there may be selection to delete organelles from sperm, once sperm are no longer free-swimming but transported via pollen tubes to the egg. More streamlined sperm may penetrate more efficiently into the egg. Although most lower plants have plastids in motile sperm (Tilney-Bassett, 1978), and most angiosperms have plastidless nonmotile sperm in pollen tubes, this hypothesis neglects the plentiful inclusion of plastids and other entities in the sperm of certain angiosperms and many gymnosperms, both of which usually have nonmotile sperm. Nevertheless, perhaps it raises an important consideration, especially for species in which male investment in the zygote is not adaptive.

3.3 SPOROGENESIS AND DOUBLE FERTILIZATION

Male and female sporogenesis exhibit important differences. The morphological sequences are generally well described but the evolutionary consequences of different patterns have been little explored. These cannot be discussed in terms of either kin selection or mate choice without considering genetic events during formation of gametophytes and gametes and those at fertilization, when the products of meiosis combine in new potential individuals.

Microsporogenesis usually produces four microspores from each mother cell (e.g., Dunbar, 1975), in contrast to the formation of most female gametophytes (Figure 3). In general, each microspore develops into a pollen grain, and thus all products of meiosis ultimately enter the male gametes, which are produced mitotically within the developing pollen grain. In angiosperms there are typically two identical male nuclei that participate in double fertilization. The number of functional sperm per pollen grain in gymnosperms, however, varies enormously: one in many Taxaceae and some *Podocarpus*, two in a variety of genera, ten to fourteen in certain species of *Cupressus*, and as many as twenty in *Microcycas* (Appendix).

In many gymnosperms additional sperm from a single pollen grain may fertilize other eggs. These additional sperm may preempt archegonia, preventing other males from fertilizing those eggs; thus they reduce intraovular female choice and enhance their own probability of acceptance. After fertilization by multiple sperm nuclei from a single pollen grain, the resulting zygotes within a gametophyte are genetically identical; which of these survives is evolutionarily inconsequential. This situation is ideal for kin-selected "sacrifice" of identical zygotes for the benefit of the survivor. It seems likely that the survivor can, in fact, use some product of its sacrificed siblings, although the identity of the fathers is not known in any instance. If kin selection op-

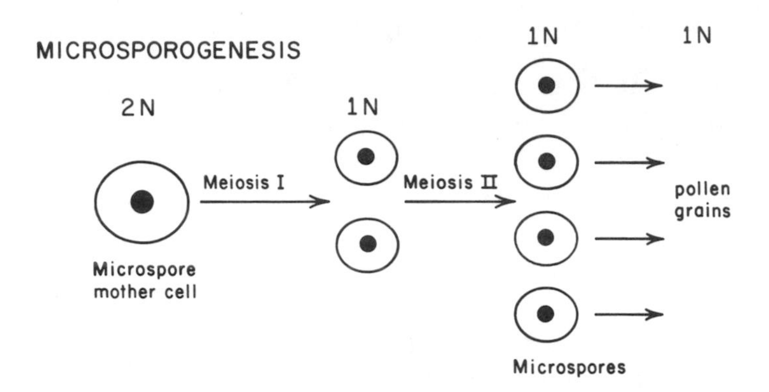

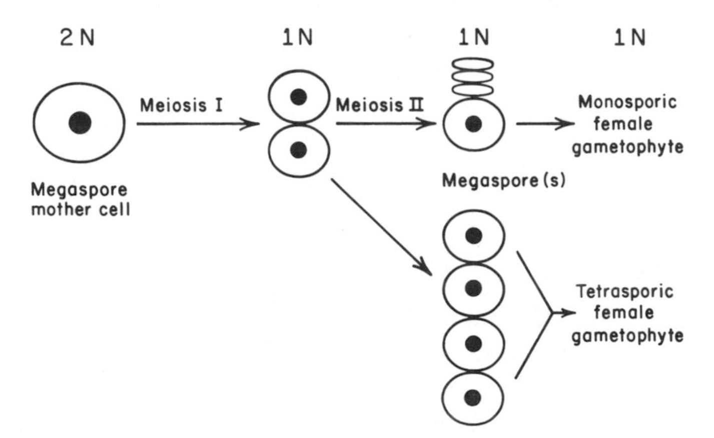

FIGURE 3. Microsporogenesis and megasporogenesis. Microspore formation usually uses all four meiotic products of the microspore mother cell in the development of four pollen grains. Megaspore formation may use only one of the meiotic products in the resulting female gametophyte or all four may contribute to a single gametophyte (see text for variants).

erates here, then multiple fertilizations in gymnosperms may be seen as a nutritional parallel to double fertilization in angiosperms (see also Favre-Duchartre, 1974).

When some of a male's gametes or zygotes are incorporated by other, successful, genetically identical embryos, the investment of the male in the failed entities might be a form of indirect paternal investment. We know that nutrition of embryos in various animals can be enhanced by consumption of extra gametes (Marshall, 1966: 258; Grell, 1967; Olive and Clark, 1978) or other, conspecific, embryos (Danill, 1932; O'Gara, 1969; Askew, 1971; Eickwort, 1973), and sibling cannibalism is not uncommon (O'Connor, 1978; Polis, 1980, 1981). Aborted embryos are absorbed by the successful one in *Callitris* of the Cupressaceae (Favre-Duchartre, 1974). Sweet (1973) mentioned transfer of nutrients from aborted ovules, so transfer within an ovule is not out of the question. The compound endosperm of the Loranthaceae is derived from endosperms of several zygotes, and in the Santalaceae, endosperm haustoria grow out and consume other ovules (Kuijt, 1969).

Female gametophytes in gymnosperms are relatively large; they bear the gametes and nourish the embryo. Nutritional products are already present before fertilization. Typically in gymnosperms, female gametophytes bear archegonia (except in *Gnetum* and *Welwitschia*) in which the eggs develop (see Maheshwari and Singh, 1967). Angiosperm female gametophytes, in contrast, are relatively small (D'Amato, 1977) and usually bear a single functional gamete. Nutrition of the developing embryo is commonly provided by endosperm, an entity resulting from the joining of a second male nucleus (in addition to the one that fertilizes the egg) with varying numbers of female "polar nuclei" (Figure 4). This curious phenomenon is called double fertilization. After penetration by the second sperm, there is but a short interval before nuclear fusion (Maheshwari, 1950). In apomictic species the endosperm may derive solely from maternal nuclei or, in some cases of pseudogamy, from double fertilization—but an asexual embryo replaces the sexual one. Queller (in press) suggests that the hypothetical "neotenous" (but see Doyle, 1978) origin of an-

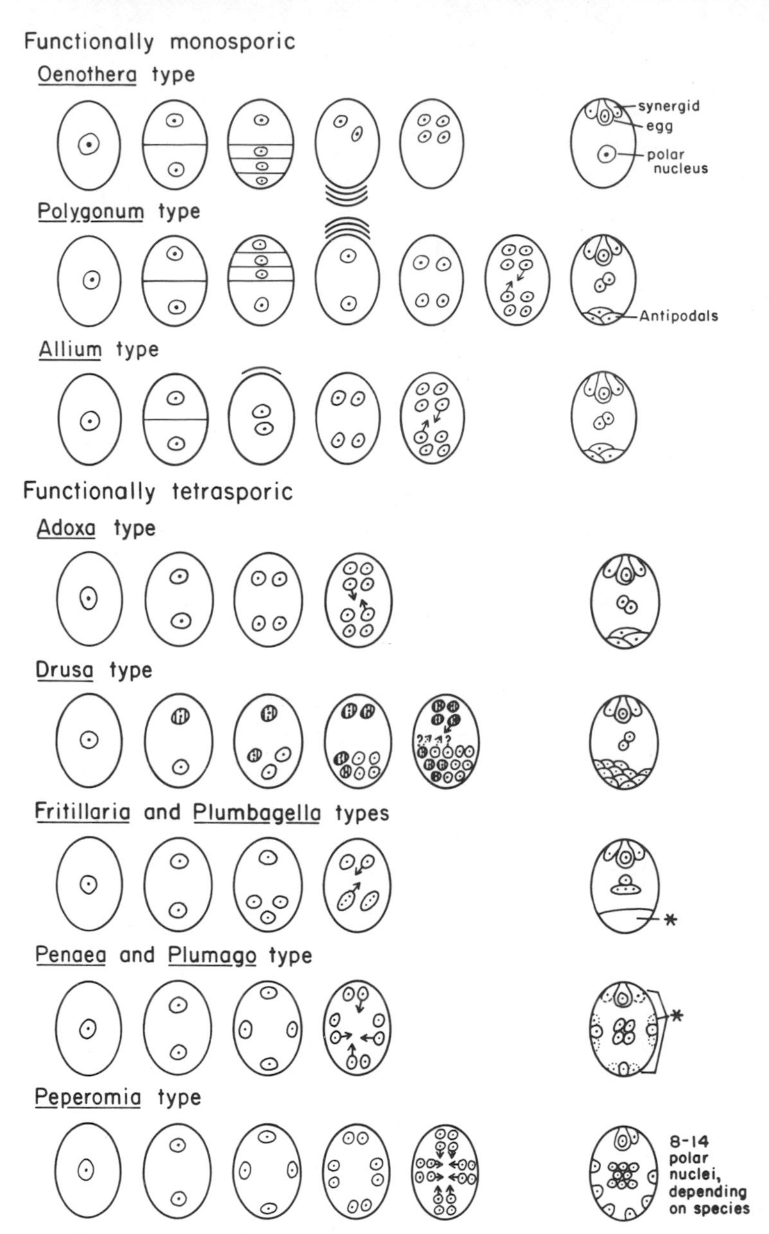

FIGURE 4. Some patterns of megasporogenesis and embryo sac formation (from Maheshwari, 1950; Johri, 1963). We have grouped these by the sources and numbers of polar nuclei rather than by the standard morphological criteria to which the names refer. For the *Penaea / Plumbago* and *Peperomia* types, Johri (1963) and Maheshwari (1950) indicated the possibility that the polar nuclei sometimes may be unequally derived from each of the four sets of nuclei, but no details were given. For the *Drusa* type the shaded nuclei at the lower end are genetically identical to those at the upper side of the (diagrammatic) embryo sac.
The asterisk indicates a variable number of nonendosperm nuclei, depending on the number of intervening mitoses.

giosperms from a gymnosperm ancestor may have been accompanied by rather undeveloped gametophytes lacking the usual stored resources, thus setting the stage for the evolution of a new mode of embryo nutrition.

From each meiosis in megasporogenesis of most seed plants there emerges a single haploid cell; this single megaspore gives rise to the female gametophyte (see, e.g., Cocucci, 1973; Foster and Gifford, 1974; Swamy and Krishnamurthy, 1975). The remaining haploid genome in these species degenerates. *Which* of the four meiotic products survives varies among taxa (Davis, 1966), but whether the observed positional effects (Figure 4) are associated with any consistent genetic patterns in the surviving haploids seems to be generally unknown. In certain *Oenothera*, however, the surviving megaspore is independent of position in the tetrad and is of a particular "genotypic class" (Grant, 1975: 231). A second type of gametophyte in angiosperms is formed by two megaspores from the same side of the first (reduction) division of meiosis (Figure 4). Because all subsequent nuclei are formed mitotically, this type of gametophyte formation ("bisporic") seems to be genetically similar to the monosporic type described above. In these functionally monosporic cases the egg and polar nuclei are identical.

Another type of gametophyte formation is called "tetrasporic," although this morphological term covers a spectrum of eventual genetic entities. Tetrasporic patterns occur in certain angiosperms and in *Gnetum* and *Welwitschia* (Waterkeyn, 1954; Martens, 1963) among the gymnosperms. Basically, all four megaspores are said to participate in formation of the female gametophyte (e.g., Sanwal, 1962; Foster and Gifford, 1974) (Figures 3 and 4), but the actual situation sometimes deviates markedly from this initial pattern. In some species (certain Asteraceae and Scrophulariaceae, and the Limnanthaceae: see Davis 1966) three megaspore nuclei eventually degenerate so that the end product is genetically similar to that from monosporic development. It is conceivable that the female sometimes has control over *which* nuclei degenerate and can exercise some choice among the haploid nuclei, or that the meiotic products compete; one or the other seems to be true in *Oenothera* (see above). Moreover, in the *Drusa*-type of embryo sac development, at the sixteen-nucleate stage there are four nuclei at one end of the sac and twelve at the other (Figure 4). Eight of the twelve are from one side of the meiotic reduction division and four are like those at the opposite end of the sac. Which of these twelve migrates to the center to contribute to endosperm formation is apparently uncertain; clearly the genetic results differ considerably, inasmuch as the resulting endosperm may have either the genetic complement of the embryo (functionally monosporic, with a double dose of one haploid maternal set) or the full maternal complement in addition to that of the sperm. In the truly (genetically) tetrasporic cases, egg and polar nuclei differ in genetic constitution.

The endosperm's function is to supply nutrition to the offspring: it cannot reproduce. Possible options are to withhold from its zygote some resources (which will be used by the mother to invest in other zygotes) or to invest "maximally" in its zygote. Withholding investment may occur through maternal (sporophyte) control or possibly through kin selection (if the relatedness of other zygotes to the endosperm is high and there is a mechanism for detecting genotypes of other zygotes). Cook (1981) suggested that the endosperm evolved through maternal manipulation, as a means of delimiting resources available to zygotes. Providing greater investment than is in the mother's best interest suggests either parent-offspring conflict or male-female conflict. Queller (in press) presents a "kin conflict" argument that parallels ours in some ways but favors an interpretation based on kin selection rather than male-female conflict, or some combination of the two, such as we develop below.

Charnov (1979) pointed out that fertilization by a second sperm raises the relatedness of the endosperm to its zygote. Fertilization produces endosperm identical to the zygote if only one female polar nucleus is involved (*Oenothera*-type). Where more than one polar nucleus is involved, the relationship between zygote and endosperm is asymmetrical, and for the kin selection rationale provided by Charnov, what really matters is the relatedness of zygote to endosperm ($r_{z(e)}$, not vice versa [$r_{e(z)}$], see section 1.1). Charnov also has suggested that the contribution of the male nucleus to the endosperm increases the competitive ability of that male in garnering resources for its zygote and preventing abortion (male-female conflict or male-male competition). If this is correct, the contribution of multiple haploid nuclei by the female to the endosperm may have evolved subsequently, as a means of decreasing the impact of double fertilization. Below we consider alternative hypothetical evolutionary origins of the endosperm and their possible significance in terms of the interests of the zygote, the mother, and the father. Because the probability of replication solely by male gametic material appears remote, we have not considered this case in detail and do not refer to it in Figure 5 or Tables 4 and 5.

Theoretically, we can envision either a single or multiple evolutionary origin for double fertilization (Figure 5). Since several types of gametophyte formation may occur within a genus (Johri, 1963), and even within species of the Asteraceae and Orchidaceae (Davis, 1966), the postulated transformations need not be very difficult. If the origin were singular, it seems most likely that the ancestral condition was monosporic, in which case the original protoendosperm (prior to the evolution of double fertilization) was likely haploid with a genetic complement identical to the unfertilized egg. The zygote and mother were equally re-

A. Monosporic Haploid Origin

F ⟶ FM ⟶ FFM - - - - ⟶ FFFM

FF'F'F'M ⟵ - - FF'F'M - - - - - ⟶ FFF'F'M

B. Multiple Origin

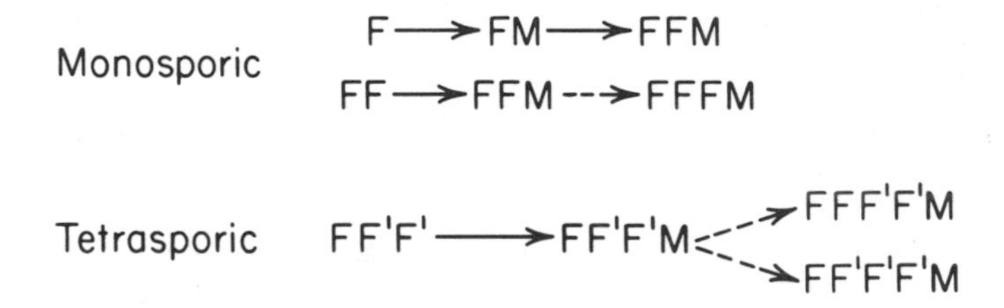

C. Monosporic Diploid Origin

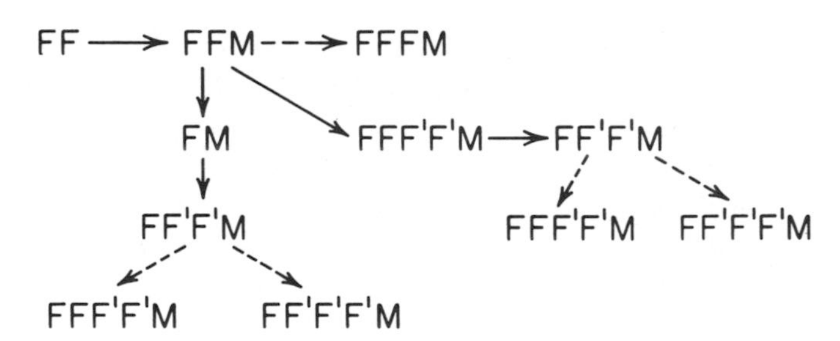

FIGURE 5. Possible evolutionary origins of the endosperm. F = maternal gametic complement; F' = maternal nongametic complement; M = paternal gametic complement. Monosporic endosperm has only the maternal gametic complement; the tetrasporic endosperm has both F and F'. Dashed arrows indicate possible further replications after the basic endosperm type has been established. See text for explanation.

lated to the protoendosperm, with other offspring of the mother related by half as much, or more if some zygotes shared the same father or if inbreeding occurred (Table 4, çol. 1). From a kin selection point of view, the protoendosperm should have had a tendency to allocate resources to its zygote, but this tendency should not have been pronounced. The relationships are proportional from a gene level point of view (Table 4, col. 1; compare odd and even rows). The evolution of double fertilization by the addition of a paternal nucleus increased the relative relatedness of the endosperm to the zygote with respect to both the mother and the mother's other offspring (Table 4, col. 3). This should have increased the endosperm's altruism toward its zygote. The shift in relationships from the genes' perspective was similar, indicating a stronger alignment with zygote as opposed to the mother. This could possibly have resulted in a reduced abortion rate through greater allocation of resources to the young.

At this point monosporic and tetrasporic endosperms must have diverged (Figure 5A). The monosporic case resulted in further replication of the gametic genome in the endosperm. This replication has no impact on kin relations that influence altruism. It does change the balance from the genes' perspective, however, with a greater proportion of genes being drawn to the mother's interests (Table 4, cols. 4, 5). Further replications could take one of three forms: female gametic contribution could differentially replicate, the male contribution might differentially replicate, or all nuclear constituents could replicate evenly. Of course, if all constituents replicate evenly, relationships do not change at either individual or gene level. Replication of female gametic material increases the "maternal" influence, whereas replication of male genetic material within the endosperm would increase "paternal" and therefore "offspring" influence. In sum, prior to double fertilization, phenomena can be understood either in terms of kin se-

TABLE 4. Summary of relationships among mother, father, endosperm, and offspring assuming a monosporic precursor of double fertilization

Relationship	Level	1 F	2 (FF)	3 FM	4 FFM	5 FFFM	6 FF′F′M	7 FFF′F′M	8 FF′F′F′M
r_{ze}	indiv.	0.50	0.50	1.00	1.00	1.00	1.00	1.00	1.00
r_{ez}	gene	1.00	1.00	1.00	1.00	1.00	0.50	0.66	0.40
$r_{♀e}$	indiv.	0.50	0.50	0.50	0.50	0.50	1.00	1.00	1.00
$r_{e♀}$	gene	1.00	1.00	0.50	0.67	0.75	0.75	0.80	0.80
$r_{♂e}$	indiv.	0.00	0.00	0.50	0.50	0.50	0.50	0.50	0.50
$r_{e♂}$	gene	0.00	0.00	0.50	0.33	0.25	0.25	0.20	0.20
r_{oe}	indiv.	0.25	0.25	0.25	0.25	0.25	0.50	0.50	0.50
r_{eo}	gene	0.50	0.50	0.25	0.33	0.38	0.38	0.40	0.40

NOTES:

Values are minimal: they assume an absence of inbreeding, and they assume that all zygotes have different fathers. Altering these assumptions does not change the direction of relationships, but it does affect their magnitude. Odd rows: the relationships of various "individuals" to the endosperm (*r* equals the fraction of genes contained by these entities also contained in the endosperm). These values reflect affiliation of the endosperm at the individual levels (pages 9-11). Even rows: the relationships of the endosperm to various entities. These values reflect relative affiliation of the genes within the endosperm to the genes contained in other entities (see pages 9-11), or gene-level selection. The second column then indicates the level of selection from the perspective of the endosperm.

F = maternal gametic complement in endosperm; *F′* = maternal nongametic complement; *M* = paternal nongametic complement; *z* = zygote; *e* = endosperm; ♀ = mother of zygote; ♂ = father of zygote; *o* = mother's other offspring (an "average" individual).

The evolutionary sequence for the diploid monosporic case is not illustrated in this figure but would encompass increases and decreases in ploidy involving columns 3, 4, 6, and 7, as illustrated in Figure 5.

lection or male-female conflict, but after double fertilization evolved, further change in ploidy can be understood only in terms of conflict between genes from the male and those from the female.

The situation is slightly different when tetrasporism evolves from the diploid endosperm (Figure 5A), for tetrasporism involves the insertion of nongametic maternal chromosomes into the endosperm. This results in restoring the original balance of maternal kin relations (with zygote and mother twice as related to endosperm as are the mother's other offspring), and at the gene level, in a greater affiliation with the mother (Table 4, col. 6). At this point kin relations are fixed and further replications of the three chromosomal sets shift the balance only at the gene level (Table 4, cols. 7 and 8). Replication of the nongametic maternal complement shifts influence to the mother; replication of the paternal complement would shift it to the zygote as does replication of the gametic complement, although the impact of replication of the gametic complement is weaker.

Now suppose that double fertilization evolved separately in tetrasporic and monosporic cases. In the monosporic case kin and gene relationships are identical, regardless of whether the ancestral condition was haploid or maternally ("bisporic") diploid (Table 4, cols. 1, 2; Figure 5B). The postulated ancestral condition in the tetrasporic case (Figure 5B; Table 5, col. 1) is interesting, for in this instance the zygote and the mother's other offspring are equally related to the endosperm, and at the gene level the maternal influence dominates. This condition seems favorable to the establishment of maternal control, allowing the mother to manipulate investment in offspring to best advantage, without "disagreement" from the endosperm. The invention of double fertilization (Table 5, col. 2), however, predictably changes these relationships to favor the zygote.

A third possible origin for double fertilization would in-

TABLE 5. Summary of relationships among mother, father endosperm, and offspring assuming a tetrasporic origin of double fertilization

Relationship	Level	FF′F′ ⟶	FF′F′M ---►	FFM
r_{ze}	indiv.	0.50	1.00	1.00
r_{ez}	gene	0.33	0.50	1.00
$r_{♀e}$	indiv.	1.00	1.00	0.50
$r_{e♀}$	gene	1.00	0.75	0.67
$r_{♂e}$	indiv.	0.00	0.50	0.50
$r_{e♂}$	gene	0.00	0.25	0.33
r_{oe}	indiv.	0.50	0.50	0.25
r_{eo}	gene	0.50	0.38	0.33

Values are minimal: they assume an absence of inbreeding, and they assume that all zygotes have different fathers. Altering these assumptions does not change the direction of relationships, but it does affect their magnitude. Odd rows: the relationships of various "individuals" to the endosperm (*r* equals the fraction of genes contained by these entities also contained in the endosperm). These values reflect affiliation of the endosperm at the individual levels (pages 9-11). Even rows: the relationships of the endosperm to various entities. These values reflect relative affiliation of the genes within the endosperm to the genes contained in other entities (see pges 9-11). The second column indicates the level of selection from the perspective of the endosperm.

F = maternal gametic complement in endosperm; *F′* = maternal nongametic complement; *M* = paternal nongametic complement; *z* = zygote; *e* = endosperm; ♀ = mother of zygote; ♂ = father of zygote; *o* = mother's other offspring (an "average" individual).

For relationships involving additional replications of *F*, *F′*, and *M*, see Table 4.

volve a triploid, tetrasporic ancestor, followed by loss of the nongametic maternal complement subsequent to the evolution of double fertilization. We view this alternative as least likely: the mother should have control over the material placed in the endosperm, and removal of nongametic complement negatively affects her control, especially with regard to the male.

None of these possibilities can be equated with what is usually considered to be the primitive condition, namely the *Polygonum* type (Figures 4 and 5C), in which the future endosperm material is diploid prior to double fertilization

(Johri, 1963; Foster and Gifford, 1974). For a triploid monosporic endosperm to evolve to a tetrasporic condition, however, both additions in nongametic maternal complement and losses of gametic complement would be required. The addition of nongametic material would benefit the mother (Table 4, cols. 4 and 7), and mechanistically this transformation could be made if following the first reduction division of meiosis all products were saved. On the other hand, the loss of the gametic complement would not be beneficial to the mother (Table 4, compare cols. 3 and 4, 6 and 7). If the tetrasporic condition did not evolve in this fashion, it is not clear in which order the addition and loss occurred (Figure 5C).

In sum, the origin of double fertilization may have arisen through kin selection involving parent-offspring conflict or through male-female conflict. We feel that parent-offspring conflict is less likely, for reasons discussed earlier (Chapter 1) and because the selection pressures acting to produce double fertilization must have concentrated on males. Within the framework of argument based on degrees of relatedness, subsequent increases in ploidy are most likely the result of male-female conflict expressed through conflict operating at the level of the gene. We have found no evidence that the male nucleus multiplies before joining the female polar nuclei, but if such a case were ever found, it would be most interesting in the possible implication for endosperm evolution.

Occasionally, the egg nucleus is fertilized by sperm from one pollen grain and the polar nuclei by sperm from another (Brink and Cooper, 1947). Effectively, one male might be "parasitizing" another by using the materials in the other's pollen to enhance the growth of his embryo.

The partial or complete elimination of endosperm and hence of functional double fertilization in certain angiosperms (e.g., Podostemaceae and most Orchidaceae: see Brink and Cooper, 1940; Poddubmayer-Arnoldi, 1960;

Davies, 1966) precludes these possibilities. Mycorrhizal nutrition coupled with the production of numerous miniscule seeds in orchids may have selected for reduction of endosperm. In addition, it may be possible that fruit production in some orchids is limited by pollen supplies (e.g., Schemske, 1980a, for data on one species; Dodson, 1962, suggests it is common among tropical orchids) so that male-male competition and male-female conflict at the level considered here are negligible.

Brink and Cooper (1940) suggested an alternative hypothesis for the origin of double fertilization, namely that the advantage may be found in heterosis and the concomitant increased "vitality" and ability of the endosperm and its embryo to draw nutrition from maternal tissues. By implication, this is a case of parent-offspring conflict. The expression of any heterotic effects should vary with the dosage of maternal genome in the endosperm, decreasing with increasing representation of the female's genes. The degree of relatedness is not central to Brink and Cooper's hypothesis; female control is exerted directly through maternal products (enzymes, etc.). Presumably, their hypothesis would predict that, when high levels of ploidy occur, the various complements (paternal, gametic, nongametic maternal in tetrasporics; paternal and gametic in monosporics) will be equally replicated. By contrast, we predict that if these replications represent further interactions between male and female and if females are able to add specific chromosome complements, there will be an underrepresentation of gametic chromosomes in polyploid tetrasporic endosperm, because this would increase the relative maternal influence on the endosperm.

An interesting hypothesis, advanced by Westoby and Rice (1982), is that the endosperm, and presumably double fertilization, evolved to minimize deceit by zygotes and thereby enhance maternal control of resource allocation. In their words:

> Why have mother plants not evolved machinery in which they obtain information . . . about the quality of offspring genotypes from the offspring tissues, but deliver the provisions to maternal tissues? Under such an arrangement, offspring would be selected to send misleading messages to the mother. A situation would evolve in which there was poor correlation between the information obtained from the offspring, and the actual vigor of the offspring genotype. . . . Offspring have had no evolutionary choice but to express the competence of their genotypes by endosperm growth. . . .

By this view it would appear that similarity between endosperm and zygote is a characteristic favored by mothers and that double fertilization is not the outcome of conflict between males and females. Instead it is a device that permits mothers to assess zygote quality and retain control over provisioning. Westoby and Rice (1982) hypothesize that increases in maternal ploidy operate to align the endosperm's interests more with those of the mother vis-à-vis the offspring. There are two sorts of problems with this hypothesis. First, the notion that the endosperm reflects zygote quality is weakened by the lack of true genetic identity between endosperm and zygote. In particular, changes in ploidy may affect rates of growth, as discussed below. Given this situation, it is unclear why growth rate of the endosperm would provide more accurate information on the zygote's quality than would growth of the zygote itself, despite differences in possible deceitfulness.

Stebbins (1976) suggested that double fertilization provides more templates for RNA synthesis and therefore facilitates rapid growth. He saw the proposed advantage as a means of shortening developmental time, apparently assuming that it would be uniformly advantageous to both sexes under all conditions. Charnov (1979), in contrast, proposed that the genetic contribution of a male to the endosperm might increase the competitive ability of that male in garnering resources for its zygote and avoiding abortion. In this view the machinery provided by the male seems to be more important then genetic constitution or kinship. If the maternal contributions to the endosperm proliferate, however, the effect of the machinery provided by the male would only be enhanced. This implies that changes in female ploidy are not a counter-measure against male input and could indeed occur in the absence of double fertilization. Westoby and Rice's, Charnov's, and Stebbins' suggestions do not seem to account for differing levels of maternal ploidy in the endosperm.

There is a further complication. Another possible consequence of double fertilization and polyploidization of the endosperm is increased nuclear and cell volumes. Price and Bachman (1976) suggested that although increased DNA content slows mitosis, this effect is more than compensated by increases in cell volume. This may have a physiological advantage in greater "succulence" due to a high ratio of protoplasm to extranuclear material, thus compensating for the putative "physiological inferiority" of the haploid gametophyte surrounded by diploid tissue (D'Amato, 1977; Cavalier-Smith, 1978). But the effect on cell size may place limits on achievable levels of ploidy if each increment of ploidy is accompanied by an increase in cell volume. The low levels of polyploidy in gymnosperms compared to angiosperms (Price *et al.*, 1974; Price, 1976) may be associated with the already high DNA content of gymnosperms (Sparrow *et al.*, 1968; Ehrendorfer, 1976) such that a further increase in cell size would slow the growth rate too much (Cavalier-Smith, 1978). Because gymnosperms can increase DNA content at least as much as angiosperms during periods of active growth (Sparrow *et al.*, 1968), however, this limitation may be less serious than supposed.

Heterochromatinization (reflecting inactivation) disrupts the link between cell and nuclear volumes and length of the cell cycle (e.g., Barlow, 1972), and thus DNA content and nuclear volume can be increased without slowing mitosis (Nagl, 1974; Nagl and Ehrendorfer, 1975). This suggests that it may be possible for males to mitigate the effects on cell volume (i.e., by being heterochromatic) without also slowing mitotic rates. A female might counter male investment that speeds development by means of mechanisms that slow development, such as increases in ploidy, giving her more time to assess male quality and/or more control over the zygote. Is it possible that chromosomes of one sex could induce heterochromatinization of those of the other sex?

A difficulty with interpreting the significance of changes in DNA content, mitotic rates, and cell volume is that most of the studies demonstrating correlations between cell and nuclear volumes with DNA content are based on interspecific comparisons (e.g., Van't Hof, 1965; Baetcke *et al.*, 1967; Bennett, 1972; Price *et al.*, 1973; Edwards and Endrizzi, 1975; Price and Bachman, 1976). Intraspecific comparisons often fail to show such correlations (Troy and Wimber, 1968; Owens and Molder, 1973; D'Amato, 1977), or else show positive correlations (Bennett and Smith, 1972; but see Gunge and Nakatomi, 1972; and below). Even intraspecifically, the potential for nongenetic developmental effects of DNA content may still be present (Willson, 1981). Though the following cases do not deal with endosperm formation, they provide hints that, even in endosperm, DNA content *might* influence rates of development. In *Spiranthes sinensis* (an orchid) facultative heterochromatinization varies with habitat and growth potential (Tanaka, 1969, in D'Amato, 1977), showing that intraspecies variability is possible. Geographic variation in DNA content is known in several conifer species (references in Price, 1976) and could be related to cell size and growth rate differences. Environmental induction of heritable (but sometimes reversible) changes in DNA content in *Linum* and its association with plant size (Durrant and Jones, 1971) also hint that intraspecific variations may be significant. The well-known shifts in DNA content with growth activity (e.g., Sparrow *et al.*, 1968, for plants; Owens and Molder, 1973; Nagl *et al.*, 1979) suggest further parallels with events in embryogeny.

Rutishauser (1956, in Nygren, 1967) has presented a fascinating case in *Ranunculus auricomus*. The second male nucleus—and sometimes a third—joins the two or three female polar nuclei, and subsequent mitoses create high ploidy levels that vary with the numbers of male and female nuclei. Such a species might be a suitable subject for exploration of relative multiplication rates of male- and female-derived nuclei and the order in which multiplication occurs. This particular species may be unsuitable because it is aposporous and pseudogamous, producing asexual embryos but requiring stimulation of the egg by the sperm (incurring one of the costs of sex without the benefits). Since the offspring are therefore descended only from the female, male-female conflict is no longer an issue for *R. auricomus*. Are there other species with such variable ploidy levels?

Nongenetic effects of double fertilization include increases in rates of synthesis and development. Further augmentation of the supply of templates and substrates by the female could enhance this effect, and unless female augmentation varies for different fathers, changes in ploidy of the maternal complement of the endosperm only enhance the male effect. Double fertilization might increase the size of the resource "sink" in the endosperm/embryo complex by altering the biochemical (including hormonal) balance in such a way that a greater supply of nutrient material is transported to that sink (Wareing and Patrick 1975). But other factors complicate the issue: increasing ploidy can often slow mitotic rates and increase cell volume, though mechanisms exist to decouple these correlations (e.g., heterochromatinization). Perhaps there can be trade-offs between the effects of template supply, cell size, and mitotic rates.

3.4 DELAY OF FERTILIZATION

Between pollination and fertilization, some time may elapse. This period of delay is characteristic of gymnosperms, in which the lapse may be as long as nineteen months (Appendix). Delay of fertilization is also common in angiosperms, although the duration of delay rarely approaches that of certain gymnosperms. Krugman *et al.* (1974) reported that for most angiosperms the typical delay is one or two days, but longer intervals are known for some species. In *Populus* the delay lasts up to three days (Winton, 1968; Fechner, 1972); in *Populus tremuloides*, probably ten to twelve days (Fechner, 1976); in *Juglans*, two to five days (Funk, 1970); in *Liquidambar styraciflua*, one to three weeks (Schmitt, 1965); in *Betula allegheniensis*, three to four weeks (Clausen, 1973); in *Citrus*, up to four weeks depending upon strain (Bacchi, 1943); in *Myrica* spp., four to seven weeks (Davis, 1966); among orchids: *Calanthe veitchii*, five weeks; *Cypripedium insigne* and *Dendrobium nobile*, ten to twelve weeks (Poddubmayer-Arnoldi, 1960); in hazel (*Corylus avellana*), four months, and in witch hazel, *Hamamelis virginiana*, about seven months (Endress, 1977); and in some *Quercus*, twelve to fourteen months (Krugman *et al.*, 1974). Here we consider possible functions for delays of short and long duration.

Short-term delays

The question of why fertilization should be delayed at all rather than being immediate or nearly so can be approached through consideration of the spatial and temporal characteristics of pollen dispersal. Most evidence indicates that pollen usually travels very short distances (but see Lanner, 1966; Thomson and Plowright, 1980; Levin, 1981). Dispersal patterns of spores and pollen generally have a leptokurtic distribution, with only small amounts of pollen traveling considerable distances (Jones and Newell, 1946; Wolfenbarger, 1946; Bateman, 1947a,b,c, 1950; Wright, 1952; Erdtman, 1954; Strand, 1956; Wang *et al.*, 1960; Silen, 1962; Whitehead, 1969; Geary, 1970; Levin and Kerster, 1974). An exceptional case, in which pollen deposition was greater at 3,000 feet from the source than at 1,000 feet was reported for *Populus nigra* by Wright (1952; and see Lanner, 1966). Some insect-pollinated species may exhibit a slight decrease of pollen deposition in the immediate vicinity of the source (Levin and Kerster, 1974), but the remainder of the distribution often resembles the leptokurtic curves. A leptokurtic pattern usually is produced whether pollen is transported by animals or by wind (Bateman, 1947a,b,c, 1950) (Figure 6). Animal pollination, however, is thought to be more efficient than wind pollination; it is associated with greater efficiency of seed set (Baker, 1963); and evidence suggests that, among angiosperms, the ratio of the number of pollen grains to the number of ovules is greater for anemophilous species (Faegri and van der Pijl, 1979). Obligate outcrosses generally produce more pollen than do partial or total selfers (Cruden, 1977).

An upper limit to pollen size may be set by the distance it can be carried (Levin and Kerster, 1974), which appears to be a more restrictive problem for anemophilous species than for entomophilous ones. Like nonbiotic particulate matter, anemophilous pollen with high terminal velocity (which is correlated with large size and relative density) settles sooner and therefore disperses over shorter distances than pollen of low terminal velocity (Wright, 1952; Pedersen *et al.*, 1961; Whitehead, 1969; Raynor *et al.*, 1970a,b; Primack, 1978). Wodehouse (1935) reported that anemophilous pollen tends to range in size from seventeen to

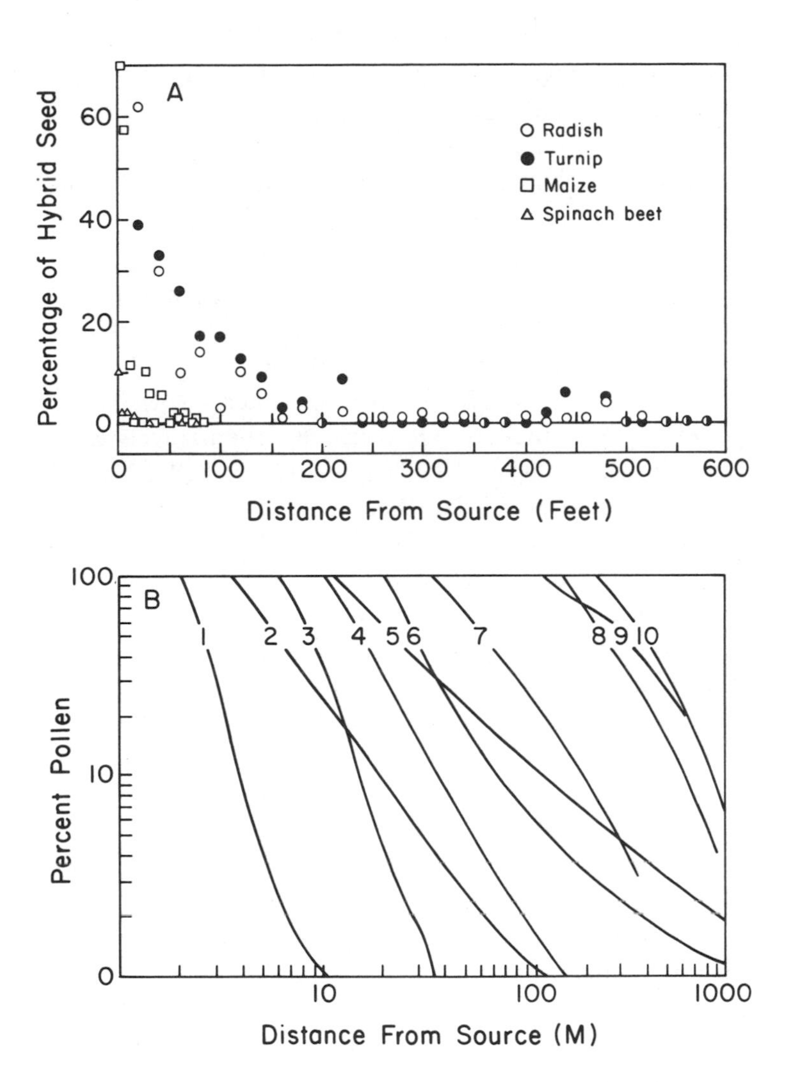

FIGURE 6. Dispersal of pollen. *A*. Percentage of hybrid seed set as a function of distance from source for several crop species. Radish and turnip are entomophilous; maize is anemophilous; beet uses both wind and insect vectors. Data from Bateman (1947a,b). *B*. Change in relative pollen concentration with distance. Base point ("100 percent pollen") is outside source. (1) *Thalictrum*; (2) pinyon pine; (3) ragweed; (4) ash; (5) slash pine; (6) atlas cedar; (7) bromegrass; (8) timothy; (9) rye; (10) cocksfoot. From Levin and Kerster (1974).

fifty-eight μ, while entomophilous pollen has a much greater range in size. Interestingly, the smallest pollen is entomophilous; Wodehouse felt that the lower limit to the size of anemophilous pollen was determined by the physical limitations of escape from a parent plant by bodies with very large surface-to-volume ratios.

Numerous factors other than size affect pollen dispersal. For both wind and animal-borne pollen, distance traveled decreases as the density of surrounding vegetation increases (Bateman, 1947a; Levin and Kerster, 1974). In zoophilous species the average distance pollen travels is affected by the behavior of pollinators—for example, whether they tend to be territorial (Wolf, 1969; Stiles, 1972), trapliners (Janzen, 1971b), or "itinerant" (Beattie, 1971; Janzen, 1971b; Ehrlich and Gilbert, 1973). For wind-pollinated species, atmospheric conditions are a parallel determinant. Wind velocity and circulation patterns affect distance traveled (Lanner, 1966; Tauber, 1967), and slow wind speed has a greater effect on heavy than on light pollen (Gloyne, 1954; Carbon, 1957). Pollen flow is restricted if the wind is not strongly directional, with a greater effect on long-distance dispersal, or if rainy weather prevails (Levin and Kerster, 1974). External morphology also influences flight distance (Stanley and Kirby, 1973).

At any given time during pollen shed, the majority of pollen accumulated on a female of a wind-pollinated species comes from sources near her (Figure 7A). Because many variables influence dispersal, the distance traveled by pollen in any population must vary between breeding seasons. The amount of pollen received from a plant's nearest neighbor should be less dependent on atmospheric conditions than should the amount received from distant plants. As a result, in good pollen and/or good dispersal years there will be larger amounts of pollen available and possibly also a greater proportion of pollen from places farther away

(Figure 7b). In poor dispersal years nearly all the pollen available will come from nearby sources (Figure 7C).

The short distance traveled by most pollen may pose a limiting factor for outcrossed, or would-be outcrossed, species. Even though female reproduction is often not limited numerically by pollen availability per se, fitness may be commonly affected by access to outcross pollen. Dispersal and receipt of "desirable" pollen may pose a particular problem for gymnosperms, which are typically wind-pol-

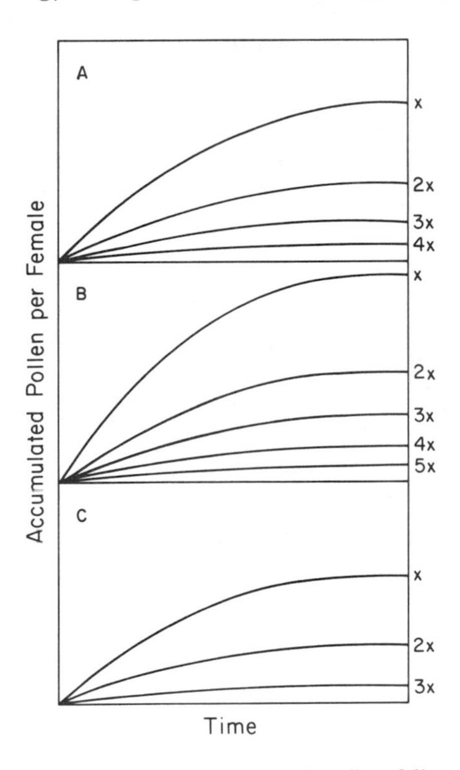

FIGURE 7. Effect of pollen abundance and quality of dispersal agent on the proportion of pollen received from different distances (multiples of *X*). Relative to an "average" year (*A*), females receive a greater proportion of pollen from distant sources in good dispersal and/or good pollen years (*B*), and a smaller proportion from distant sources in poor years (*C*). Total amount of pollen received also varies.

linated (except possibly for certain cycads: see Faegri and van der Pijl, 1979; but see Giddy, 1974) and whose pollen has a high terminal velocity (Levin and Kerster, 1974). Anemophilous angiosperm pollen is reported to travel farther than anemophilous gymnosperm pollen (Stanley and Kirby, 1973).

Another relevant difference between gymnosperms and angiosperms is that gymnosperms have quite restricted abilities to assess pollen quality prezygotically, whereas angiosperms often have some ability to differentiate among intraspecific pollens from different sources (see Section 3.1). A combined consequence of lack of prezygotic mate selection and short average dispersal distance of pollen is that a majority of fertilizations take place between related individuals, especially in gymnosperms. If moderate-to-high levels of inbreeding are advantageous, this poses no problem to females, but if inbreeding is disadvantageous, females that somehow reduce the proportion of inbred zygotes or offspring formed will be selected for. To the extent that the ability of angiosperms to detect and "neutralize" pollen of close relatives is less than perfect, angiosperms may experience similar, but less intense, selection for post-zygotic means of discriminating against inbred offspring.

Delaying fertilization may be a female tactic to increase relative degree of outbreeding or to acquire the greatest number of potential mates. In the absence of any delay, pollen from nearby males has a substantial competitive advantage. To see this, consider an example in which pollen arrives at a female from two distances, X and Y. Y is twice as far away as X. This means that in any given time interval, pollen from X is over twice as likely to arrive as pollen from Y (Figure 8); the actual relationship is an inverse square function. Given that X and Y types have equal viability and are equal in total abundance, X pollen will have more than two times as many *independent* opportunities to fertilize a given ovule as will Y. Let the probability (p) of successful fertilization for both X and Y types be 50% when they arrive at previously unfertilized ovules. In order to determine the probability of fertilization by a pollen type, we need to know the order of arrival. As a simple approximation when $Y = 2X$, we will assume that two X's typically arrive before each Y. In this case the relative probability of fertilization by X pollen is 0.75 $(1 - p^2)$. As the distance between X and Y increases, the probability that a given ovule will be fertilized by closer pollen before the more distant pollen first arrives rapidly approaches certainty (curve 1 in Figure 8). If four X's arrive before a Y, for example, the probability of fertilization prior to Y's arrival is 0.94 $(1 - p^4)$.

By delaying fertilization until all pollen has had maximal opportunity to arrive (or at least until the receptive surface is entirely covered), females in effect increase competition among pollen from nearby sources and promote the pos-

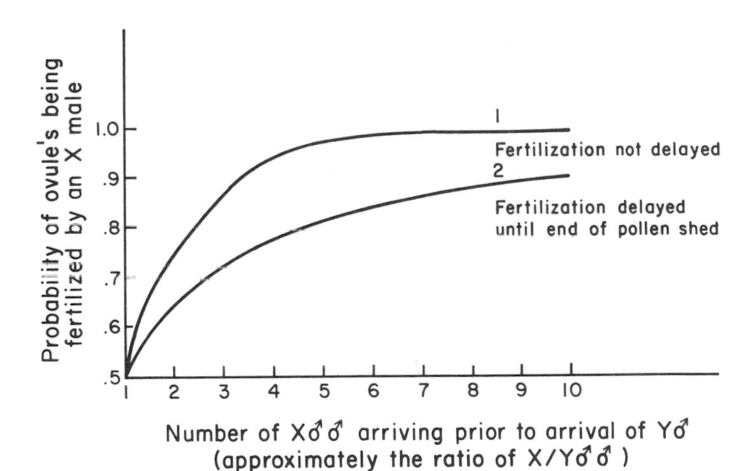

FIGURE 8. Probability of being fertilized by an "X" male depending upon whether or not fertilization is delayed. X and Y indicate distance of male parents from the female. Upper curve: fertilization is not delayed. Each pollen grain that arrives has a probability of 0.5 of successfully fertilizing a previously unfertilized ovule. Lower curve: fertilization is delayed until end of pollen shed. Probability of fertilization is proportional to relative abundance of pollen types.

sibility of fertilization by males farther away. Fertilization by faraway males is by no means ensured through delaying; up to a point the relative probability of such fertilization should be inversely proportional to relative distance of two pollen sources (curve 2 in Figure 8). When pollen is abundant, early arriving pollen may still have a disproportionate advantage, since receptive surfaces or pollen chambers may be filled (e.g., Sarvas, 1968; Stern and Roche, 1974).

In sum, females may alter the spectrum of potential mates simply by varying the timing of fertilization from no delay (selection for inbreeding, assuming neighbors are relatives) to a duration slightly beyond the termination of pollen shed (selection for the greatest degree of outbreeding or variety of mates). Beyond the end of pollen shed further delay does not increase the range of available mates, for pollen does not continue to accumulate, and other factors must be responsible for delays of greater length (see below). The absence of delay in fertilization should be expected only in species that preferentially inbreed and have prezygotic mechanisms for evaluating pollen quality.

As mentioned above, angiosperms may benefit from delaying fertilization if prezygotic selection is imperfect. Another function of delays in angiosperms may be the accumulation of maximum amounts of pollen on stigmatic surfaces in order to permit females to effect varying degrees of selectivity (see section on female choice). That is, females may benefit from accepting lower-quality pollen when superior pollen is unavailable. By delaying fertilization, females may give high-quality pollen the maximum opportunity to arrive before they have to "decide" to accept or reject pollen of lesser quality.

Long-term delays

Short-term delays in fertilization may be viewed as a female reproductive tactic designed to increase the overall quality of a season's brood. By this reasoning, after the end of pollen-shed females should permit fertilization to take place. Then females would begin a process of zygote selection, aborting certain categories of zygotes and retaining preferred ones. If the process of abortion is somewhat staggered, however, for reasons suggested below, males may benefit from interfering with the abortion process by delaying fertilization even further. We suggest that this tactic is advantageous when circumstances favor females being less selective toward ovules fertilized late.

Abortion may be performed sequentially for several reasons. First of all, females may maximize their control over zygote fate by staggering the times of fertilization of different eggs. The evaluation of embryo quality is probably a relatively time-consuming process, because it involves chemical integration of information over some portion of the plant. This may limit the scale at which a female can make simultaneous decisions and favor sequential decision making. Furthermore, after fertilization fathers may be selected to accelerate embryo development, because the benefits of abortion to the female decrease as development proceeds (since less future investment is required). While the simultaneous evaluation of all embryos sometimes might be desirable in terms of making precise comparisons, it may be impracticable.

There are several more proximate reasons for staggering fertilization and/or abortion. If the precise timing of seed production is not critical, staggered production may evolve because it permits periodic reassessment of resource availability, an idea also suggested by D. A. Goldman (personal communication). This creates selection for females to make rapid, economical decisions concerning abortion. In order to do so, they may have to restrict the number of decisions being made simultaneously, prolonging the period over which fertilizations occur. By this means, females may adjust their abortion rate in response to improved estimates of resources. If the degree of competition among pollen varies at different ovules, fertilization may be effected at different rates. Different portions of the plant may have different resources available, permitting different development rates. In addition, some species have protracted flowering periods (Heinrich, 1975, 1979; Frankie, 1975; Levin, 1979), producing a small number of flowers at a time. When this is the case, it may be reasonable to view offspring as belonging to successive, overlapping clutches. In this case the situation resembles one in which flowering periods are much shorter but fertilization is delayed for one breeding season or more (see below).

Whether or not fertilizations are staggered, females may have to stagger abortion, simply because of the temporal constraints placed on the process of embryo evaluation. In some species fruiting is less synchronous than flowering (e.g., *Hybanthus prunifolius, Randia armata*: see Augspurger, 1981a). Fruiting is sometimes more synchronous, however, and although fruits that were initiated early in the season have a higher probability of maturation than later ones, a pattern of delayed fruit maturation reduces their advantage (e.g., *Cassia fasciculata*: see Lee, 1980).

Recall that the decision to abort a particular embryo should be based on the abundance of embryos of higher quality. When females whose reproduction is primarily resource-limited practice a sequential abortion process, natural selection may favor a sliding scale of zygote selectivity. That is, at the beginning of the selection process females may abort all but the best embryos but decrease selectivity as the process continues. A primary reason for this hypothesized procedure is that females are unlikely to be able to assess the absolute abundance of superior embryos prior to onset of abortion. As the process proceeds, however, it is reasonable to expect that information increases, so that females can lower their selectivity as the number already aborted approaches the fraction that is advantageous to abort.

The logic of this selection procedure can be best appreciated by considering the consequences of adopting the opposite strategy, namely to begin by being less selective and to increase selectivity during the abortion process. A consequence in this case would be that females would spare lower-quality ovules without knowledge about the availability of higher-quality ones. In years in which the proportion of higher-quality pollen is moderate to high, females would invest more in offspring of lower quality than they would have if they had been highly selective initially.

We can imagine circumstances that would select against females practicing a decreasing scale of selectivity. Most likely this practice would be disadvantageous when females are sometimes pollen limited, because they would run the risk of aborting offspring they could not replace. If females were often, but not always, limited by pollen availability, then we would expect to see increasing selectivity when pollen becomes abundant, a phenomenon reported for the trumpet creeper (*Campsis radicans*: see Bertin, 1982c). When pollen availability fluctuates considerably among breeding seasons, it should be advantageous to detect relative pollen availability prior to the onset of abortion. The possibility of prezygotic female sensitivity toward pollen availability has been suggested for *Pinus sylvestris* by Sarvas (1962), who indicated that the duration of female receptivity may be related to the amount of pollen received in the pollen chamber. In *Pinus sylvestris* and in *Picea abies* the average size of the pollen chamber is large enough to accommodate at least enough pollen to fertilize all the archegonia in an ovule (Sarvas, 1962, 1968). Even when pollen is abundant, some ovules fail to become fertilized, owing to wind directionality during the interval of pollen shed (Sarvas, 1968). For this reason it would be quite advantageous for females of species not typically "limited" by pollen availability to be nevertheless sensitive to the lack of pollen at some fraction of ovules.

If females practiced sequential abortion using a sliding scale of selectivity, ovules fertilized late would be aborted less frequently than those fertilized early. Under these circumstances, males would benefit from delaying fertilization. Initially, strategic delays might be for very short intervals, measured in hours or days if the normal duration of fertilization is brief. Given a sliding scale of selectivity, however, over time male-male competition could generate directional selection to delay fertilization for longer and longer intervals. Reports that fertilization in *Ginkgo biloba* often occurs *after* ovules are shed from the mother (Eames, 1955; Favre-Ducharte, 1958) are tantalizing. Nevertheless, plants of some species preferentially mature fruits resulting from early fertilization (e.g., *Yucca whipplei*, see Aker and Udovic 1981) and, in *Campsis radicans*, nonfavored pollen donors are more successful in early pollinations than in late ones (Bertin 1982c).

Males encounter an additional and conflicting selection pressure for early fertilization in gymnosperm species with multiple archegonia per ovule. Males whose pollen delayed fertilization would not prevail over males with nondelaying pollen arriving at the same ovule. An evolutionary bind is present in that early fertilization is undesirable if it leads to probable abortion but may be necessary to ensure fertilization at all. Males may counter this problem by suppressing, perhaps chemically, the activity of other pollen arriving at an ovule. (This apparently can happen interspecifically [Sukhada and Jayachandra, 1980]; can it occur intraspecifically also?) Pollen might then become active and fertilize an archegonium when timing is strategic to avoid abortion, or when its ability to suppress the activity of others declines, whichever occurs first. The ability of pollen to delay its own fertilization will be dependent on its ability to suppress the activity of other pollen, because if other pollen becomes active, fertilization and possible abortion will ensue. Thus selection will continue to favor males whose pollen is able to suppress others the longest, for that pollen is both most likely to fertilize and to fertilize last. "Warfare" among pollen therefore reinforces and should extend the period of fertilization delay. The ultimate length of the delay should be affected by environmental and physiological constraints operating on females. There are other constraints as well: greatly extended delays in the timing of fertilization would postpone the age of first reproduction and reduce the total number of seed crops produced in the female's lifetime.

Females might be unable to combat male-initiated (or male-lengthened) delays. Any females that evolved a preference for nondelaying males would simultaneously have to lower their selectivity for other traits. That is, they would have to *not* abort the offspring of a male they would otherwise abort in order to have an effective preference for early fertilizing males. Females with preferences for early fertilizers may also be selected against for essentially Fisherian (Fisher, 1958) reasons: they produce sons whose pollen does not delay and so become aborted by their mates. Thus, as a result of a variety of selection pressures acting on both males and females, fertilization could be delayed for extended intervals of time.

When delays evolve in excess of the length of one breeding season, a new possibility emerges. At this point females may use information concerning the subsequent year's pollen crop to assess the relative quality of the previous year's and therefore to alter the abortion rate levied against zygotes from the previous year's pollen crop. Two kinds of information may assist females in evaluating the relative quality of sequential pollen crops: direct information on the quality of individual pollen grains (in angiosperms) and information on the overall pollen crop (in both angiosperms and gymnosperms). In good dispersal years pollen should be more available than in poor dispersal years, and for outcrossed species the quality of pollen received should be substantially higher. Females of species lacking the ability to detect pollen quality prezygotically may, nevertheless, be able to evaluate the relative size of pollen crops, as can females that have good prezygotic assessment. Once a relative evaluation is made, females could adjust their investment in the earlier year's progeny, allocating greater or lesser resources to it, depending on the pollen received in the subsequent year. If year-to-year evaluations are possible, this could contribute to the evolution of mast-fruiting patterns.

Short and very long delays may therefore represent female tactics to maximize the quality of offspring that receive further investment following ovule production. Delays of any duration after the period of pollen shed may result from selection on males to lower the incidence of abortion of their pollen. The frequency and duration of extensive fertilization delays should depend upon the importance of abortion as a female reproductive tactic. The more sophisticated prezygotic detection becomes, the less important postzygotic control should be. The fact that delayed fertilization is much less extensive in angiosperms, which possess more sophisticated prezygotic detection abilities, is consistent with the hypotheses developed here.

Of course, not all instances of delayed fertilization necessarily reflect the outcome of conflicting male and female interests. Various other factors, particularly ecological considerations, may affect the timing of pollen dispersal, fertilization, and seed maturation. For example, short-term delays (those less than one year) could result if different times of year proved suitable for pollen dispersal and seed maturation. Plants might evolve more extensive delays in response to a shortening of the reproductive season. When the time needed for seed maturation is long relative to the season available for both pollination and maturation, it may be advantageous to practice clutch overlap (Burley, 1979) and mature the previous season's crop concurrently with pollination of the present year's brood. In neither of these examples, however, would it appear necessary for fertilization itself to be delayed; rather, the zygotes could simply become dormant sometime after fertilization until conditions appropriate for maturation occurred. Nevertheless, when delayed *maturation* is advantageous for reasons such as those presented above, delayed *fertilization* may subsequently evolve in response to the pressures of sexual competition outlined in this section.

3.5 POLYEMBRYONY

Polyembryony is the production of more than one embryo per gametophyte. We will consider three types: simple, cleavage, and adventitious (see reviews by Webber, 1940; Wardlaw, 1955; Maheshwari and Sachar, 1963; Nygren, 1967).

Simple polyembryony (SPE) will here refer to the presence in each female gametophyte of more than one egg (Figure 9) and the potential for each egg, if fertilized, to produce an embryo. SPE is common in many ferns (Buchholz, 1922) and other lower plants (Rickett, 1923; Showalter, 1926, 1927a,b; Chatterjee and Mohan Ram, 1968). It is widespread in gymnosperms (Appendix), where typically only one embryo survives in the mature seed (Kozlowski, 1971). The occasional maturation of more than one embryo, especially common in some species, would be worth investigating carefully.

Cleavage polyembryony (CPE) refers to the production of usually genetically identical, multiple zygotes from a single fertilized egg (Figure 10). Characteristic of certain invertebrate animals and some armadillos (Patterson, 1927; Silvestri, 1937; Eisenberg, 1966), and occurring in some angiosperms (e.g., Davis, 1966; Jong, 1976), CPE is extraordinarily well developed in the gymnosperms. In fact, many authors remark that it is one of the most represent-

GYMNOSPERMS:

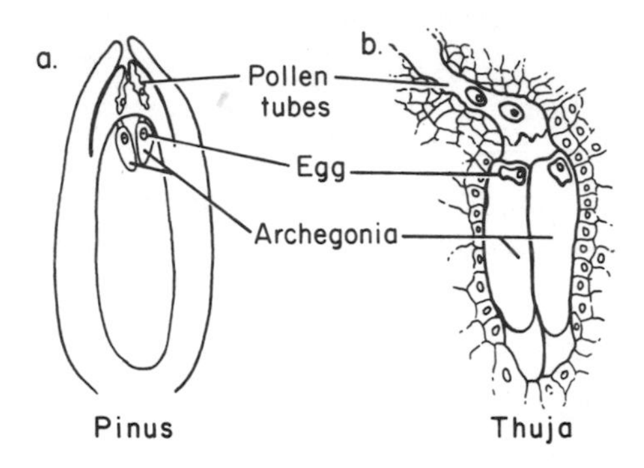

ANGIOSPERMS:

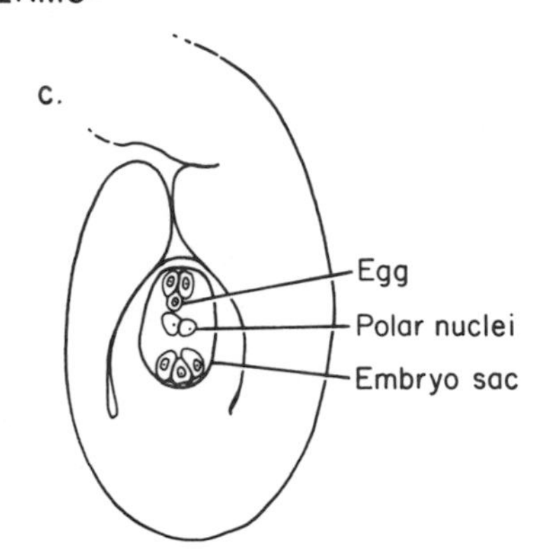

FIGURE 9. Ovules. *A*. Multiple archegonia, each having separate entry points for sperm. *B*. Close-up of an archegonial complex with a common entry point. *C*. For contrast, an angiosperm ovule with a single embryo sac is shown.

ative and typical gymnosperm features (e.g., Chamberlain, 1935). In the gymnosperms there are two major forms of CPE, which we discriminate by the location of cleavage activity. Primary CPE occurs in the terminal embryonic-initial cell of the developing zygote, which may cleave to form several dozen embryos. Secondary CPE refers to the development of embryos from the rosette or suspensor cells

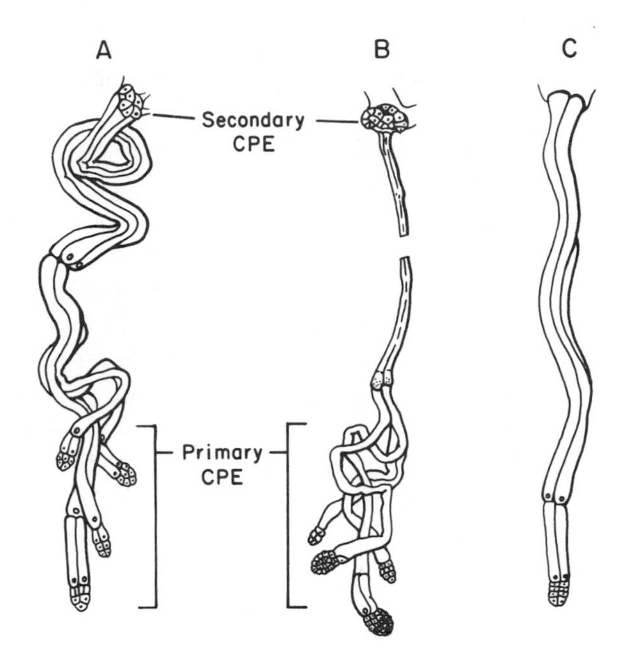

Figure 10. Embryos of three conifers. *A. Pinus laricio (nigra)*; *B. Cedrus libani*, both showing CPE; and *C. Pseudotsuga taxifolia*, which has no CPE (from Buchholz, 1931, in Wardlaw, 1955).

of the developing zygote. In some cases several hundred cleavage embryos, of both sorts, may be initiated, but typically only one persists to become the definitive embryo in that ovule. In general, it seems that secondary cleavage embryos almost never become the definitive embryo. A basic evolutionary question is why CPE and SPE occur, if some embryos have no chance from the start and so many are assured of an early demise (e.g., Buchholz, 1926; Ching and Simak, 1969).

Asexual embryos are rather common among angiosperms (Maheshwari, 1950, Nygren, 1967) but are rare in gymnosperms (Dogra, 1966a). Some occur as the result of embryogeny in an unreduced, unfertilized egg, but usually only one embryo per ovule develops to maturity (deWet and Stalker, 1974), so we can neglect these here. Less common, but relevant to our discussion, are adventitious embryos that develop directly from maternal sporophytic tissues; several such embryos often mature (Webber, 1940; Jong, 1976), resulting in adventitious polyembryony (APE). This condition is reported from a number of genera and is particularly common in *Citrus* and *Opuntia* (Nygren, 1967).

These three forms of polyembryony can be distinguished genetically. In SPE the eggs all arise from the same gametophyte and thus possess identical maternal haploid genomes. The zygotes may or may not have different fathers, depending on the number of sperm per pollen tube (in some gymnosperms this can reach sixteen or twenty; Appendix) and the source(s) of the pollen grains. In CPE the nuclear genomes of all embryos are identical. And in APE (as we use the term), embryos have only the maternal genome (or an aneuploid variant thereof [deWet and Stalker 1974]).

SPE means that a potentially varied array of zygote genotypes within each ovule presents the female with the material for choice among them: by whatever means, she may be able to choose the best combinations from the array. This device may permit the elimination of inbred zygotes, at least those with homozygous lethals, without loss of the ovule, as suggested by forest geneticists (Bramlett and Popham, 1971; Stern and Roche, 1974; Lindgren, 1975; Sorensen, 1982). It may also reduce the frequency of other undesirable or less compatible combinations (Sorensen, 1982).

A parallel phenomenon may occur in multiovulate carpels in angiosperms. Female reproductive organs may contain various number of ovules; in some species all can develop, in others all but one are aborted. For instance, six ovules typically develop in *Quercus*, of which five abort early (including some that get fertilized) and only one normally matures (Mogensen, 1975); in *Triplochiton scleroxylon*, there are eight to ten ovules per carpel but usually only one develops—two on some individuals (N. Jones, 1976). Of six ovules in *Hopea odorata*, five degenerate but, curiously, polyembryony is well developed (Jong, 1976). Only one of four ovules per fruit in *Betula allegheniensis* develops into a seed (Clausen, 1973). Furthermore, multiple embryo sacs per ovule are reportedly common in the Hamamelidaceae and Coryloidae (Endress, 1977), although presumably only one normally participates in the formation of the seed. Still other examples are available (Bradley and Crane, 1965; Davis, 1966; Kuijt, 1969; Godley, 1979; Player, 1979). Regular patterns of abortion appear to occur regardless of levels of pollen receipt or nutrition, so the question arises as to why these "supernumerary" ovules are produced in the first place.

The occasional presence of more than one female gametophyte per ovule in certain gymnosperms (Appendix) and in some angiosperms (Johri, 1963; Davis, 1966) may also be considered here. The case of certain loranthaceous mistletoes is particularly intriguing: in some species several gametophytes (embryo sacs) grow into the style, even reaching to the stigma. Fertilization occurs in the tips of the sacs and the resulting proembryos race back down the style along the growth path of the sac. The slower ones die along the way and eventually only one survives (Kuijt, 1969; Davis, 1966). Are the zygotes being tested in some way?

Although historical explanations may be relevant, perhaps these cases of ovule abortion in angiosperms represent evolutionary equivalents of simple polyembryony, that is, a mechanism by which females choose among different fathers and/or different combinations of zygote genomes. The packaging of multiple pollen grains in clusters (Willson, 1979; Kress, 1981) may then be viewed as an analog of the gymnosperm trait of multiple sperm per pollen grain. The formation of multiple pollen tubes per pollen grain in the Malvaceae (Davis, 1966) could act to block access by other males or increase growth rates by augmenting uptake from the style. An additional variable in the angiosperms, however, is the likelihood that the maternal haploid genomes are also different in different ovules.

Because all cleavage embryos are identical (nuclearly at least), there is little basis for choice among them. Rather, a possible *raison d'être* may lie in differential growth rates; more rapid growth of the embryonic unit may decrease the probability of abortion. Sarvas (1962) noted that, for *Pinus sylvestris*, the embryo from the most advanced archegonium has the most vigorous CPE, which may further retard development of embryonic units from other archegonia. Although Buchholz (1929) argued that a single large embryonic unit should be able to prevail over a set of subdivided cleavage embryos, Doyle and Brennan (1971) stated that in fact slender cleavage embryos grow faster. No data were presented to support this statement, but it seems plausible that increased surface area, with concomitantly increased potential for uptake of materials and for surface-active enzymes of divers functions, and perhaps increased templates and machinery for synthesis, might indeed foster more rapid embryonic growth than possible for a single embryo. Even secondary cleavage embryos could contribute to this end.

If indeed CPE is a means of increasing embryo growth rates, it is a tactic perhaps most likely to be controlled by males. If a male can somehow increase development rate of his offspring, he forces the female to commit resources to his zygotes. Because of this investment and because future investment in these zygotes is then reduced (Dawkins and Carlisle, 1976; Boucher, 1977), females should be less likely to abort them. On the other hand, it is possible to view CPE as regulated by females—as a means of inducing competitive growth and choosing among the offspring of different fathers (divide and see who conquers!). In either case we should find that degrees of CPE vary intraspecifically, in part as a function of the intensity of male-male competition and the severity of female rejection. If it is solely a male tactic, the quantity and quality of competing pollen should not affect the tendency of a pollen donor to attempt to induce CPE. If females regulate CPE, however, the levels of CPE may vary with the variety of pollen sources and levels of pollen abundance for each female.

Williams (1975) suggested that cleavage polyembryony in armadillos may be a tactic of the offspring to garner more maternal resources and to increase their own fitness (in proportion to the number of identical sibs produced). From the mother's perspective, this is disadvantageous, because she encounters a potential cost of parthenogenesis (identical offspring) without the usual benefit of transmitting *all* her genes. In short, Williams suggests that the offspring have won the parent-offspring conflict. Whatever the case for armadillos, such reasoning is unlikely to pertain to plants, in which only one cleavage embryo survives.

APE is an intriguing condition, requiring the explanation of both asexual embryo formation from maternal tissue and the frequency of plural embryos. A discussion of the entire subject of asexual reproduction is well beyond the scope of this essay. We may note, however, that studies of animal parthenogenesis (references in Willson, 1981) may suggest some ecological and evolutionary parallels: by and large, parthenogenetic animals are found in regions with lower environmental diversity, under conditions in which the costs of mate-finding or sexual reproducton are particularly high, or in disturbed circumstances (but see Lynch MS). The unusual case of gynogenesis in animals, in which sperm are required to stimulate parthenogenetic development of an embryo, may find an analog in pseudogamous plants, in which the events of pollination or fertilization may trigger APE (Webber, 1940). The balance between sexual and asexual reproduction in certain facultatively apomictic species is subject to environmental control (e.g., Saran and deWet, 1976), but the adaptive value of the shifts in sexuality is apparently unstudied. According to Frost (1938), seedlings from adventitious embryos of *Citrus* are "superior" (under what conditions?) to seedlings derived from selfing and from certain crosses but inferior or equal to widely outcrossed seedlings.

Formation of *multiple*, asexual embryos may result from several factors. The great prevalence of APE in such cultivated genera as *Citrus* and *Mangifera*, in which cultivars often have a hybrid origin, opens the possibility that artificial breeding and cultivation programs have contributed to the development of APE (see also Zatyko *et al.*, 1975). In the cultivated soybean homozygosity of a male-sterile gene is associated not only with male sterility but also with abnormality of female gametophytes and increased polyembryony (Cutter and Bingham, 1977). APE is also reported for several species and subspecies of tropical trees, however (Dipterocarpaceae; Kaur *et al.*, 1978). Perhaps more interesting in an evolutionary sense are the observations that the frequency of multiple embryos in *Citrus* varies with the nutritional status of the parent (Traub, 1936) and in *Mangifera* with the variety or strain and location of growth (Juliano, 1934, 1937; Singh, 1960). Furusato (1953a,b; 1960 in Nygren, 1967) reported that different hybrids of *Citrus* have different tendencies to produce adventitious embryos and that the trait of mono- versus polyembryony is controlled by a single pair of alleles. Evidence for genetic control means that the trait can respond directly to selection.

Maturation of plural embryos raises an unstudied problem; neither costs nor consequences are understood. *Citrus* polyembryos suffer a mortality rate proportional to the number of embryos (references in Webber, 1940), so even though several embryos may reach maturity, they do so at the cost of partial development of several others. Although multiple seedlings from a single seed or fruit may all be capable of establishment and growth if experimentally separated (e.g., *Acer saccharum*: see Carl and Yawney 1972), their ability to do so when establishing naturally in very close proximity to each other is unknown. Up to seven seedlings of *Mangifera indica* may emerge from a single seed (Singh, 1960), and multiple seedlings occur in certain tropical trees (Kaur *et al.*, 1978); their fate has not been studied. Interestingly, certain individuals are apparently especially prone to producing multiseeded fruits, even when single-seeded fruits are usual (Buchholz, 1941a). The potential for severe competition among the several emerging individuals exists, but their genetic identity (in APE) may permit some form of kin-selected "cooperation." The number of fathers of multiple seeds in a dispersal unit might be manipulated by either males or females as a means of regulating sib competition and/or cooperation (Kress, 1981). A high, positive correlation of size among neighbors in clonal plantations but a lower, more variable, and often negative, correlation of size among neighbors in plantations propagated from genetically variable seeds has been interpreted as indicating a lack of competition among neighbors when they were genetically identical (Sakai *et al.*, 1968). Differences in competitive response require that seedlings have some way of physiologically assessing their neighbors' identities, however, or that such differences are linked in some way to an expectation of neighborhood composition. Apparently, neither of these has been documented.

3.6 SUMMARY OF CHAPTERS TWO AND THREE

In these two chapters we have provided the following kinds of evidence and arguments that support the proposed model of mate competition and mate choice in seed plants:

1) Resource limitation of fruit/seed production appears to be common. Limitations due to pollen quantity seem to be less common; those related to pollen quality are not well understood, because the relative costs and benefits of in-

breeding are seldom known and may indeed vary with resource levels. Individual differences in degrees of self-fertility and genetic compatibilities provide material for relevant experimentation and, when heritable, a basis for natural selection.

2) Possible cellular mechanisms of zygote control by males include earlier time of allelic activation (no evidence as yet), especially of alleles related to rates of embryo development. If such evidence is found, we predict that different males will bear alleles that activate at different times relative to females, that is, that such temporal patterns are not species-fixed. Other possible mechanisms (again no direct evidence) involve *B* chromosomes, whose activity, behavior, effects, and variability suggest a basis for mate choice, and gene amplification in gametes. Cytoplasm from male gametes of most gymnosperms and certain angiosperms enters the egg at fertilization. If it is not quickly inactivated or destroyed, it may interact with female RNA or proteins (no direct evidence specifically for the interaction of male material with female, but zygotes can inactivate maternal substances), male organelles or substrates could contribute to zygote metabolism (many indications that this is not only possible but likely), and/or male cytoplasmic genes may affect embryo growth (there is evidence that in some species male transmission occurs and varies with the nuclear genotype but apparently none regarding the consequences for the zygote).

3) Genetic relationships between embryo, endosperm, and mother in angiosperms vary with the mode of megasporogenesis and subsequent multiplication of female nuclei; these might provide bases for kin-selected traits and the degree of maternal control over her offspring. Double fertilization is more likely to have evolved monophyletically through functional monosporism or to have had multiple origins than it is to have had a monophyletic origin in tetrasporism. If endosperm nuclei multiply differentially,

we expect the maternal nuclear complement in the endosperm to be replicated more than the paternal complement in monosporic polyploid endosperms, and the nongametic maternal complement to be differentially replicated in tetrasporic polyploid endosperms. Multiple sperm production in certain gymnosperms may permit preemption of eggs by the fertilizing males and subsequent kin-selected "altruism" among the resultant zygotes.

4) Delayed fertilization is common among gymnosperms and some angiosperms. Short-term delays can be seen as female tactics to increase the proportion of usable pollen received from distant males, but such delays only function until the end of the period of pollen arrival. At that time, if zygote selection by females is gradual and if females are less selective of ovules fertilized late, then further delays of fertilization may reflect male tactics. Extremely long delays could again be female-induced if they permit choices among the potential offspring of more than one season.

5) SPE in gymnosperms may be a means of female choice among zygotes from different fathers. Automatic abortion of a fixed proportion of ovules in angiosperms may have a similar function. CPE, best developed in gymnosperms, may be a mechanism to increase zygote growth rates. Perhaps it is a male tactic to reduce the risk of abortion, or conceivably it is a female tactic to assess male quality. The function of APE in angiosperms requires explanation of both parthenogenetic reproduction and multiple seedlings, neither of which is well understood.

CHAPTER FOUR

Consequences of Prezygotic and Postzygotic Choice

The purpose of this chapter is to provide an overview of how selection for mate choice by females and selection on males for being "chosen" or "not rejected" may have contributed to the evolution of certain morphological and life history traits in gymnosperms and angiosperms (Figure 11). Our starting point is a rather striking difference in the basic mechanism of mate choice by the two groups of higher plants: gymnosperms appear to have very weakly developed prezygotic detection abilities, whereas among angiosperms such detection abilities appear to be well developed. As a result, selection on female function in gymnosperms has been to perfect abortion techniques, whereas selection on male function has been to prevent abortion. In the angiosperms the situation is more complex, as both pre- and postzygotic selection may be practiced to a considerable extent. We make no attempt to investigate or explain the phylogenetic origins of this difference but assume that it is an ancient one with ramifications in later evolution.

4.1 GYMNOSPERMS

When mate selection is primarily or exclusively postzygotic, a large fraction of zygotes formed are candidates for abortion. The intensity of selection on females for an efficient system of abortion is therefore high. Two considerations of importance under these circumstances are that abortion be performed as early as possible, to minimize the

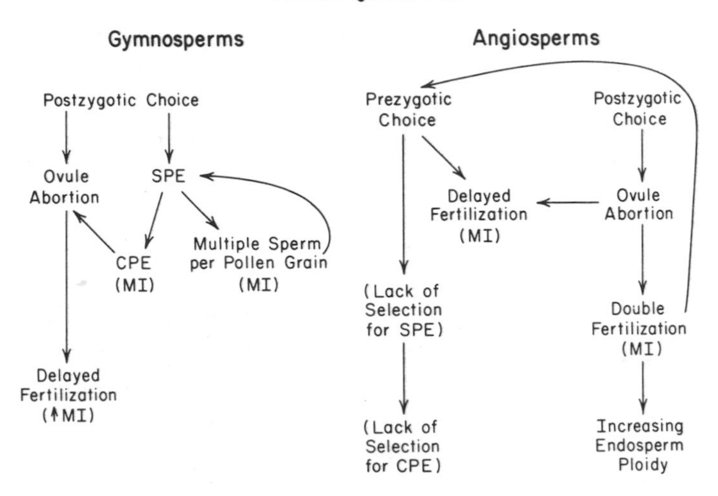

FIGURE 11. Consequences of mate choice mechanisms in gymnosperms and angiosperms. SPE = simple polyembryony; CPE = cleavage polyembryony; MI = mating investment.

cost, and that the abortion rate be adjusted to reflect optimum brood size. These two pressures conflict somewhat, since evaluation of zygote quality followed by determination of an optimal level of abortion at a plant-wide scale requires integration that is most probably time-consuming.

A method that allows local, and hence rapid, detection of zygote quality is simple polyembryony. Given adequate levels of pollination and fertilization and a diversity of pollen donors, SPE can produce an array of zygotes within each ovule. A female may therefore select the best zygote at each ovule shortly after fertilization, improving the overall quality of her "brood" with minimal expense and without any initial decrease in brood size. Within each ovule the maternal genetic complement is constant, so females utilizing SPE simultaneously select for mate quality and complementarity, either directly through chemical destruction of some zygotes or indirectly through offspring competition, in addition to passive elimination of homozygous lethals. Indirect selection seems likely to be important, since there is no *a priori* reason to expect that females can detect offspring quality soon after fertilization while being insensitive to pollen quality prior to fertilization. SPE provides a mechanism that facilitates indirect selection through embryo competition. Possibly the developing embryo can efficiently metabolize its less fortunate sibs, making the selection process inexpensive.

Although SPE will improve the quality of offspring in each ovule, variability in quality within the brood will undoubtedly remain, and selection may favor augmenting SPE with more finely tuned and probably more expensive techniques. Other factors, such as changing resource levels, may also dictate further brood reductions. Nevertheless, the potential utility of SPE to efficiently eliminate the worst zygotes in plants that lack prezygotic rejection abilities seems substantial, and in fact SPE is a nearly universal gymnosperm trait (Appendix).

The degree of mate selection made possible by SPE is a function of the number of archegonia per ovule, which varies between, and in many cases within, species (Appendix). Factors that may affect the number produced include the energetic efficiency of ever-increasing degrees of SPE, the number of pollen grains that can be expected to reach any ovule, and the variability in quality of pollen at each ovule. Responses by males may decrease the effectiveness of SPE in mate choice: for example, they might begin to produce multiple sperm per pollen grain (Figure 11), a trait documented for a number of species (Appendix). The effect of this innovation is to increase the probability of a particular male's offspring being the successful embryo within any given ovule. Females might respond in turn by increasing the number of archegonia per ovule, thereby fostering directional selection for increasing numbers of sperm per pollen grain (Figure 11; Appendix). The expense undertaken by males to secure an ovule, probably with consequent reduction in quantity of pollen produced, can be viewed as mating investment (MI in Figure 11). That growth factors such as calcium can overcome the "pollen population effect" (Brewbaker and Kwack, 1963) suggests that males might use such substances, in addition to nutrients, as MI.

The production of multiple sperm per pollen grain is one possible mechanism of countering female-initiated SPE. Another possibility is cleavage polyembryony. In CPE the zygote divides, forming multiple smaller embryos that may assimilate nutrients at a faster rate than would a single larger embryo. "Altruism" among these perfectly related entities should be evident, with all united in the "goal" of securing the ovule against competing embryos fathered by other males. Thus we view the primary function of CPE as a response to SPE: identical cleavage embryos work to sequester sufficient resources to prevail over embryos of other genotypes in that ovule. Once CPE has evolved, it may gain an additional function during competition between ovules. In this case successful cleavage embryos may have differentially harnessed resources so that females evaluate their quality partly on the basis of future investment required to mature them. At this level (but not within ovules) CPE may act as a mechanism to prevent abortion and therefore could be functional even if SPE were lost (presumably secondarily).

Evidence that CPE is common when males typically package sufficient sperm per pollen grain to fertilize all archegonia in an ovule would not support our hypothesis that CPE is a male reproductive tactic. Available data indicate that the average number of archegonia per ovule is considerably greater than the average number of sperm per pollen tube, however (Appendix). This condition may persist because the allocation of multiple sperm per pollen grain places selection pressure on females to increase the degree of SPE. Multiple sperm per pollen grain and CPE may be complementary mechanisms for securing ovules.

Both mechanisms probably require some resource allocation by males and therefore may be considered mating investment (MI in Figure 11). Neither CPE nor multiple sperm might be expected to occur in the absence of SPE. Most but not all evidence is consistent with this expectation (see Appendix).

Earlier (section 3.5) we suggested that CPE might be a female-initiated tactic of indirectly selecting mates by enhancing competition among developing embryos. While we cannot exclude this possibility completely, it does seem less likely for two reasons: 1) there seems to be no basis for assuming that the rate at which cleaved embryos grow is a better test of offspring (and thereby mate) quality than the rate of growth of embryos not displaying cleavage and 2) the effectiveness of SPE would be diminished if CPE enhanced the competitive advantage of early arriving pollen by permitting such pollen to sequester a large share of resources.

A potentially major problem in mate choice that SPE does not alleviate concerns the timing of arrival of pollen. In the absence of delayed fertilization, zygotes fathered by pollen that arrives early probably experience a competitive advantage over the products of pollen that arrives late (within the same ovule), but the mating quality of males that donate early pollen is not necessarily superior to other males and may be lower if early pollen tends to come from nearby sources. A female's average brood quality is lowered to the extent that suboptimal pollen has a competitive advantage. Under these circumstances choice of mate is facilitated through synchronizing fertilization events by delaying fertilization throughout the period of pollen shed. This tendency should be exaggerated if CPE enhances the advantage of early arrivers by increasing the utilization of resources provided by the female and thereby making such resources less available to late arrivers. A likely consequence of delayed fertilization is that males are forced to increase the energy supplied to pollen (MI) in order to maintain their embryo's ability to cleave and grow rapidly.

Through SPE females can accumulate multiple potential fathers for most or all of their offspring. Through delayed fertilization females increase the competition among pollen donated by these fathers, resulting in enhanced brood quality. They may then further improve offspring quality, at the expense of quantity, through selective ovule abortion. When females abort using a "declining scale of selectivity" (section 3.1), males can benefit from suppressing fertilization (at the local level) for longer periods since females will be less discriminating towards late-fertilizing males. This situation opens the possibility for directional selection for longer and longer delays, whose ultimate length should be determined by ecological constraints acting on the female. Beyond delays of one breeding season, females may participate in causing further delays if they can use information concerning environmental suitability and the quality or quantity of a subsequent year's pollen crop to determine the optimal investment to make in the previous year's brood.

Several preadaptations might have facilitated the evolution of extensive fertilization delays. (1) The longer the delay, the more investment males must place in pollen to insure sperm viability and zygote vigor. Selection pressure for MI probably preceded the evolution of long delays, however (Figure 11); delays therefore necessitate increased investment in each pollen grain but not the invention of new adaptations. (2) Following the invention of SPE, there was increased potential for males to benefit from interfering with the potency of other males' pollen, including the suppression of fertilizing ability. (3) Once the ability to delay had been perfected (we hypothesize by females), males might have been able to capitalize on the mechanism, perhaps amplifying it for their own ends.

Through male-male competition and female choice directional selection for increasing the energetic value of pollen may be continuous. Other factors probably set the ultimate constraints on MI by male plants. The more energy that is allocated to each pollen grain, the fewer grains can be made (for the same total allocation). Also, of course, weight increases with increased reserves. Increased weight shortens dispersal distance, decreasing the already low probability that any one grain reaches an appropriate receptive site. It is therefore unlikely that male plants could contribute large amounts of investment to each offspring. Nevertheless, even small amounts may have substantial impact on female RS and may be favored despite the cost in pollen quantity and dispersibility.

4.2 ANGIOSPERMS

Unlike gymnosperms, angiosperms are known to have well developed prezygotic, as well as postzygotic, mechanisms of mate choice (Figure 11). In general, there should be an inverse relationship between the degree to which species utilize one set of mechanisms versus another; however, even species with the most sophisticated prezygotic detection abilities may employ postzygotic selection to some extent, particularly in situations in which environmental quality fluctuates, resulting in changes in optimal brood size.

One gymnosperm trait does not appear very functional for angiosperms. If our interpretations of the functions of simple polyembryony are correct, SPE effects roughly the same results as (or acts in place of) prezygotic selection. Lacking SPE, *embryos* need not compete within ovules, so there is less selection for CPE or the production of many sperm per pollen grain.

On the other hand, delayed—perhaps synchronous—fertilization may evolve when prezygotic selection is practiced, because females need information about pollen quality and availability in order to make good mate choices.

The extent to which long delays in fertilization occur should depend on the degree to which ovule abortion is practiced. Where it is common, males may benefit from delaying fertilization to protect their offspring from abortion. Where it is very rare, selection pressure on males to delay is absent; females could still use delays in excess of the length of one breeding season to determine brood size from the previous year's pollen crop, however.

Double fertilization may have evolved through selection on males to lower the probability of abortion of their offspring. If double fertilization is particularly effective in ensuring nutritional investment in offspring that females would otherwise abort, selection may enhance female ability to select mates prezygotically (Figure 11). Further increases in endosperm ploidy may reflect conflict over control of offspring fate.

Where delayed fertilization occurs and/or selective ovule abortion is common, pollen may contain male MI comparable to that postulated for gymnosperms. The presence of strong prezygotic mechanisms may tend to decrease MI, however, for the following reason: where prezygotic selection is minimal, males that fail to make sufficient MI will fail to secure ovules, regardless of their genetic quality. On the other hand, females in many angiosperms presumably can select mates prezygotically, to some degree, on the basis of genetic quality and/or complementarity. Pollen of high quality or complementarity containing little MI should be preferred over lower quality pollen containing more MI. Mate quality and and certainly complementarity will vary among individuals, but males cannot predict where specific pollen will land. The more MI apportioned to pollen, the heavier it becomes and, if it is wind-borne, the shorter the distance it will travel. If wind-pollinated females prefer outcrossed pollen, then lighter pollen traveling a farther distance should have a mating advantage. At the other end of the continuum, however, low-quality pollen containing high MI will have a better chance of fertilizing females than pollen of similar quality containing little MI; and for any given quality, females should prefer pollen of those males that contribute investment. Since most pollen does not travel great distances, the probability of landing on nearby females is high and competition for paternity at those sites may necessitate contributing MI (and possibly PI).

These considerations suggest that in species where the problem of dispersal is small or pollen competition is low, minimal investment in pollen should be made, but where dispersal is very limited in range or pollen competition is high, investment-rich pollen should be made. Where both selection pressures exist (for example, where the probability of long-distance dispersal is variable but the potential payoff significant), one possible outcome is disruptive selection for pollen size, that is, the evolution of dimorphic or polymorphic male characteristics (e.g., Gadgil, 1972; Cade, 1979; Hamilton, 1979). Additional adaptations in pollen morphology or in dispersal mechanisms may occur so that some pollen is designed to travel relatively long distances whereas other pollen is not. Most reports of pollen polymorphism are for heterostylous species, however (Ganders, 1979), where distance traveled may affect pollen size only indirectly (through a trade-off with number). Pollen dimorphism has been reported for one nonheterostylous species, *Silene alba*; male plants produce one of two pollen types, which vary in several morphological aspects including pollen diameter. It appears that the dimorphism is genetically controlled (McNeill and Crompton, 1978).

We can summarize this hypothetical scheme in the following way: the variety of sexual strategies utilized by angiosperms and gymnosperms derives from the extent to which females use pre- and postzygotic mate choice. The dependence of gymnosperms on postzygotic choice makes SPE advantageous. This, in turn, creates selection for CPE and multiple sperm per pollen grain. Multiple sperm augments selection for SPE. CPE forces females to rely on the other postzygotic mechanism, ovule abortion. Ovule abortion occurs in both groups and may lead to the evolution of delayed fertilization; prezygotic choice in angiosperms reinforces this tendency. CPE, multiple sperm, and double fertilization probably require MI, while delayed fertilization is likely to enhance MI. Female angiosperms may have responded to double fertilization by increasing endosperm ploidy and by strengthening their prezygotic capacity for choice. Double fertilization in angiosperms may also be a response to ovule abortion and may represent MI by males.

CHAPTER FIVE

Avenues for Exploration

Our objective has been to explore the potential role of mate choice in influencing characteristics of higher plants and to suggest some mechanisms by which mate choice may be effected. We have surveyed a large literature, but there is supporting evidence for only a few aspects of our model of mate choice. On the other hand, we have not been able to find evidence that invalidates the model. By and large, biologists have not asked questions or gathered data in ways that permit evaluation of our ideas. In what follows, we present a number of hypotheses, some original, others not, that center on the testable aspects of our model. We also pose some questions where hypothesizing is not yet appropriate.

To facilitate understanding of how our hypotheses are interrelated, we present a "theory map" (Figure 12) of sexual selection in plants. The map emphasizes mate choice but necessarily includes aspects of male "competition." (The term "competition" here includes both direct interactions among males as well as indirect competition that results from male-female interactions such as the attempt by males to prevent abortion of their young.) Besides showing how hypotheses are related to one another, the theory map indicates the centrality of the various hypotheses. Hypotheses 1, 2, and 3, which we term "first-order hypotheses" are most fundamental to a theory of mate choice in plants. Some hypotheses (e.g., H_{13}, H_{14}, H_{29}) have implications for a series of others. Still others, of differing levels of importance to the theoretical structure indicated by the map, are particularly interesting for other reasons: e.g., H_{36}, because male-male interference has been so little contemplated in plants and has many ramifications; and H_{48}, because of the possibly conflicting selection pressures on male traits so basic to their reproductive success.

In the following list hypotheses that are directly testable by experimental means are marked with an asterisk. A queried asterisk (?*) indicates that the hypothesis is potentially testable by experimental manipulation, pending the discovery of particular techniques (such as a biochemical means of altering the period of delay between pollination and fertilization or of inducing CPE). An asterisk in parentheses marks hypotheses that are testable by correlations across a number of species or populations. For succinctness, closely related hypotheses of parallel form sometimes are presented as compound hypotheses (e.g., H_6, H_7, H_{12}, H_{34}, H_{43}).

First-order hypotheses

*H_1. Male plants (or the male function of hermaphrodites) compete for mating opportunities, and female plants (or female function) discriminate among males.

*H_2. The greater the usual degree of outbreeding in a population, the more intense will be female choice and male-male competition.

*H_3. Female choice and male-male competition are more intense when female reproductive success is resouce-limited rather than pollen-limited.

Related hypotheses

*H_4. Resource limitation of female RS is more common than pollen limitation.

*H_5. Females can enhance the intensity of male-male competition to facilitate mating decisions; and they can cue in on signals used in male competition and use them to make mating decisions.

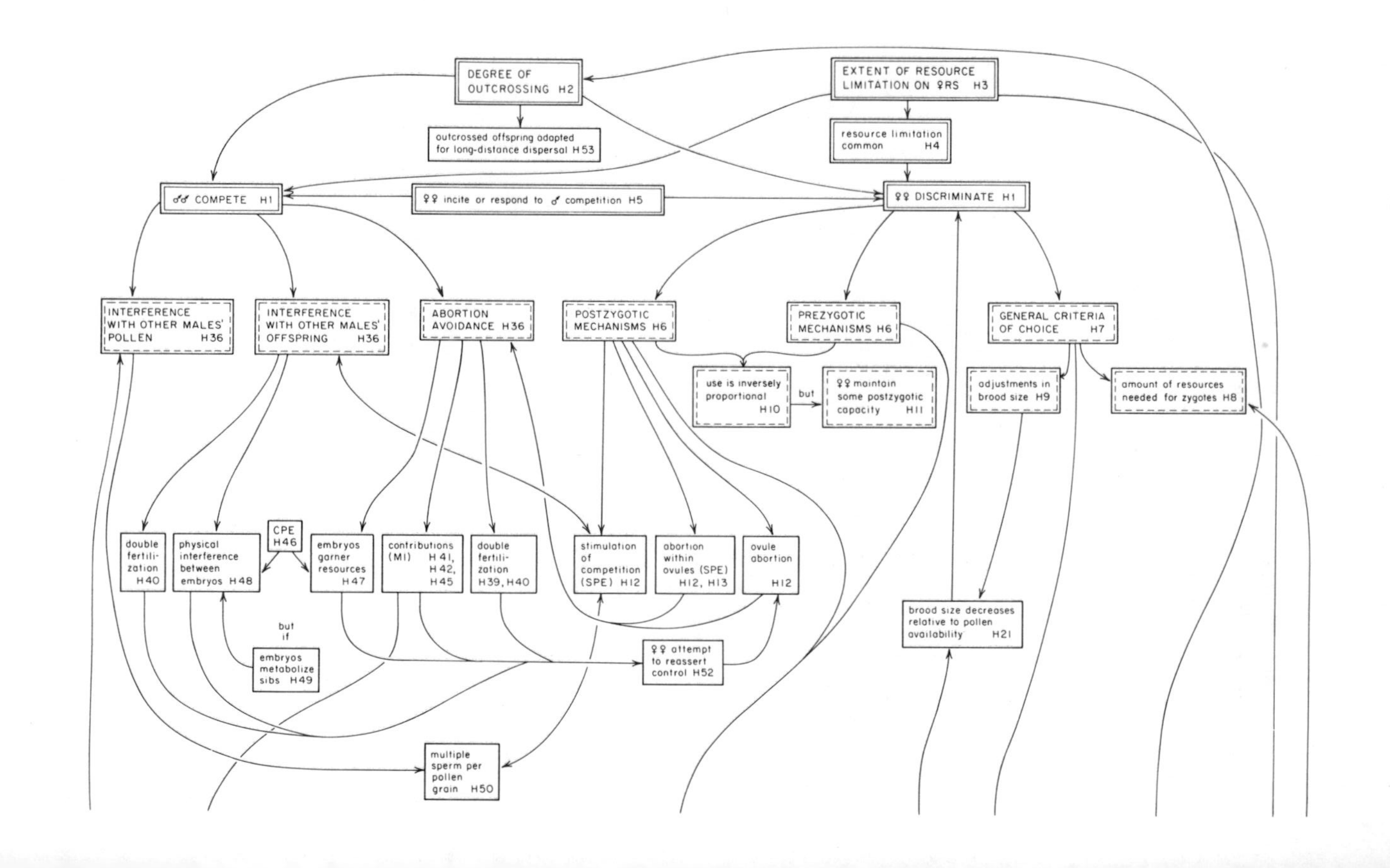
DEGREE OF OUTCROSSING H2
EXTENT OF RESOURCE LIMITATION ON ♀RS H3
outcrossed offspring adapted for long-distance dispersal H53
resource limitation common H4
♂♂ COMPETE H1
♀♀ incite or respond to ♂ competition H5
♀♀ DISCRIMINATE H1
INTERFERENCE WITH OTHER MALES' POLLEN H36
INTERFERENCE WITH OTHER MALES' OFFSPRING H36
ABORTION AVOIDANCE H36
POSTZYGOTIC MECHANISMS H6
PREZYGOTIC MECHANISMS H6
GENERAL CRITERIA OF CHOICE H7
use is inversely proportional H10
but
♀♀ maintain some postzygotic capacity H11
adjustments in brood size H9
amount of resources needed for zygotes H8
double fertili-zation H40
physical interference between embryos H48
CPE H46
embryos garner resources H47
contributions (MI) H41, H42, H45
double fertili-zation H39, H40
stimulation of competition (SPE) H12
abortion within ovules (SPE) H12, H13
ovule abortion H12
brood size decreases relative to pollen availability H21
but if
embryos metabolize sibs H49
♀♀ attempt to reassert control H52
multiple sperm per pollen grain H50

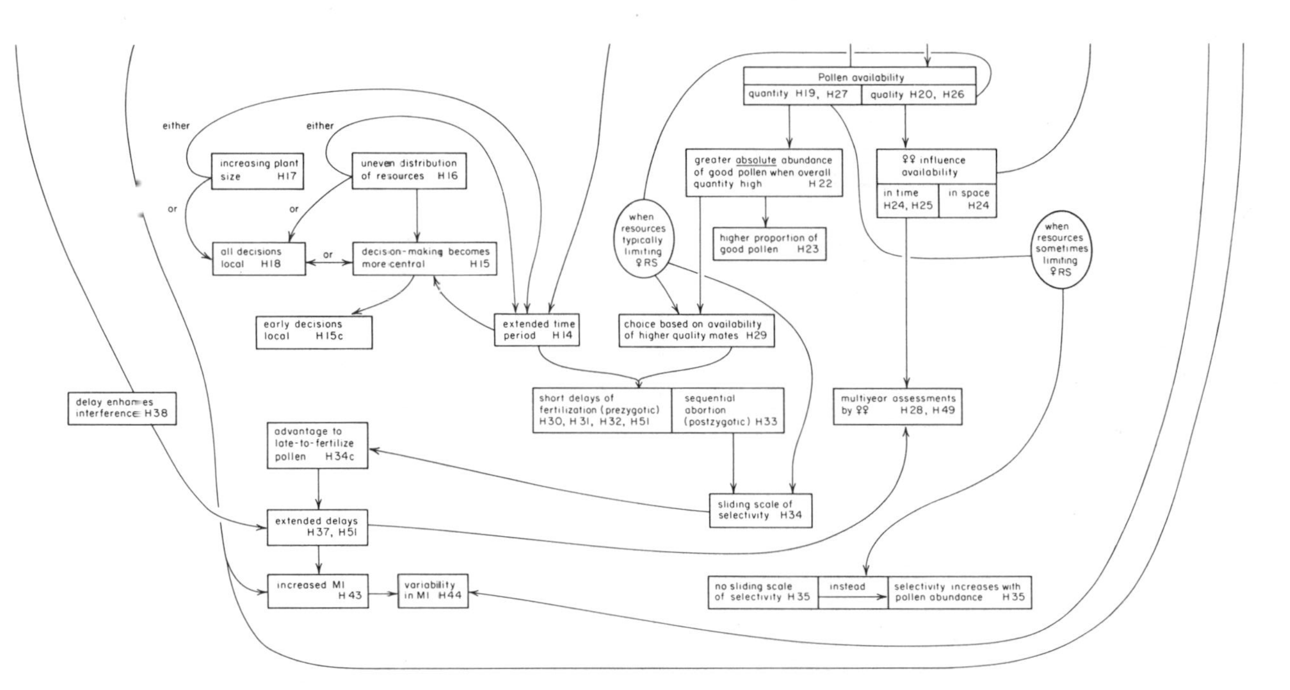

FIGURE 12. Theory map of sexual selection in plants. First-order hypotheses (all caps) and related ones (lower case) are in double boxes. Second-order hypotheses (all caps) and related ones (lower case) are in boxes with broken lines. Third-order hypotheses are in single boxes. Statements in circles are limiting conditions for hypotheses. Only relatively direct interactions among the hypotheses are illustrated here (by unidirectional or bidirectional arrows). Numbers in boxes correspond to those in the list of hypotheses in the text.

Hypotheses concerning mechanisms and tactics of female choice

Second-order hypotheses

*H_6. Mate choice can occur before and/or after fertilization.

*H_7. Female plants choose mates on the basis of genetic quality, genetic complementarity, and/or mating investment.

Related hypotheses

*H_8. Maternal decisions to abort particular embryos are influenced by the amount of further investment required to mature them.

*H_9. Females can vary their brood size in response to resource availability.

*H_{10}. The greater the development of prezygotic mate choice, the less the development of postzygotic choice.

*H_{11}. Because resource levels and pollen availability are seldom entirely predictable, females usually retain some capacity for postzygotic mate choice.

Third-order hypotheses

*H_{12}. Abortion of ovules, abortion within ovules, and stimulation of male competition are postzygotic mechanisms of mate choice.

*H_{13}. Females use simple polyembryony to make postzygotic mating decisions without aborting ovules.

COROLLARY: Simple polyembryony is a functional substitute for prezygotic selection.

*H_{14}. Both pre- and postzygotic mate choices occur over time because of the difficulty of integrating information over large segments of a plant.

*H_{15}. As the process of choice continues, decision making tends to become more centralized.

COROLLARY: Early mating decisions are likely to be made at a local level within the plant.

*H_{16}. An uneven distribution of resources within a plant restricts the ability to integrate information and make mating decisions on a large scale (i.e., over an entire plant).

*H_{17}. As plant size increases, integration of information becomes more difficult and mate choice should be more sequential.

*H_{18}. Alternatively, when resources are distributed unevenly or the plant is large, choices will be made primarily at a local level (not over the whole plant).

*H_{19}. Females vary selectivity as a function of the quantity of available pollen.

*H_{20}. Females vary selectivity as a function of the quality of available pollen.

*H_{21}. Female selectivity increases as brood size decreases in relation to pollen availability.

*H_{22}. The absolute abundance of available, high quality pollen is greater when the overall quantity of available pollen is high.

*H_{23}. The relative proportion of available high quality pollen increases under conditions favorable to pollen dispersal.

(*)H_{24}. Females influence pollen availability using spatial and/or temporal mechanisms.

(*)H_{25}. The range of available mates is increased by increasing the size of a female's receptive surface.

*H_{26}. Angiosperm females evaluate the quality of individual pollen grains.

*H_{27}. Both angiosperm and gymnosperm females evaluate the size of the pollen crop.

*H_{28}. Females subject to long delays of fertilization can use information about a certain year's pollen crop to assess the relative quality of the previous year's

crop and alter the abortion rate levied against the earlier zygotes.

*H_{29}. When resources are typically limiting, females evaluate acceptability of mates of any given quality in terms of the availability of higher quality mates.

(*)H_{30}. Short delays of fertilization increase the number of mates accessible to a female.

(*)H_{31}. Short delays of fertilization increase the possible degree of outcrossing; selection on females for inbreeding favors lack of delay.

(*)H_{32}. Short delays of fertilization permit females to make prezygotic mating decisions that reflect the availability of high quality mates.

*H_{33}. Sequential abortion permits females to make postzygotic mating decisions that reflect the availability of high quality males.

*H_{34}. When resources are typically limiting and zygote abortion sequential, female selectivity declines during the period of abortion, resulting in a "sliding scale" of selectivity.

COROLLARY: When abortion is sequential, pollen is more likely to be accepted when it is late to fertilize.

*H_{35}. When female reproductive success is commonly pollen-limited, females should not use a sliding scale of selectivity; instead they should increase selectivity when pollen becomes abundant.

Hypotheses concerning mechanisms and tactics of male competition

Second-order hypotheses

*H_{36}. Males compete for mating opportunities through interference with other males' pollen, interference with other males' offspring, and prevention of abortion.

Third-order hypotheses

(*)H_{37}. Sequential abortion and a sliding scale of selectivity favor long delays of fertilization, because pollen that fertilizes last is least likely to be aborted.

(*)H_{38}. Male-initiated delays of fertilization result in intense pollen competition.

H_{39}. Double fertilization reduces the risk of abortion.

H_{40}. Double fertilization increases the resources provided to a zygote, in one of the following four ways:

- *H_{40A}. As the relatedness of the zygote to the endosperm increases relative to the relatedness of the mother to the endosperm, the endosperm provides increased amounts of resources to the zygote. This occurs at maternal expense. (Partly testable.)
- ?*H_{40B}. Male genes in the endosperm "attempt" to activate transfer of maximal amounts of resources to the embryo (regardless of relatedness as in H_{40A}).
- ?*H_{40C}. For mechanical reasons related to DNA content and cell volume, *any* increase in ploidy, regardless of the genetic complement involved, increases the ability of the endosperm to provide for the zygote.
- *H_{40D}. Double fertilization increases the vitality of the endosperm through heterosis.

*H_{41}. Male contributions to zygotes can influence zygote growth rates and hence the risk of abortion.

(*)H_{42}. Male mating investment is positively correlated with the degree of reliance on postzygotic mate choice.

(*)H_{43}. Delayed fertilization favors increased male MI.

(*)H_{44}. Wind-pollinated species subject to sexual selection will show greater variability in MI and pollen size than those not subject to sexual selection (this occurs because of conflicting selection pressures for pollen dispersal and making investments to secure mates).

*H_{45}. Additional means by which males influence growth and abortion include early activation of paternal genes, *B* chromosomes, gene amplification, blockage of female RNA or proteins, cytoplasmic genes, and contributing organelles and materials for metabolism of the zygote.

?*H_{46}. Cleavage polyembryony increases the growth rates of embryos.

*H_{47}. Within an ovule embryos with high levels of CPE can garner sufficient resources such that abortion by the female is unprofitable (the additional time and/or resources necessary to mature the embryo diminish).

?*H_{48}. Alternatively, high levels of CPE allow physical interference of rapidly growing zygotes with slower growing ones.

*H_{49}. Surviving embryos (produced by both SPE and CPE) can metabolize their "outcompeted" sibs (acceptance of H_{49} favors H_{48} over H_{47}).

*H_{50}. Production of multiple sperm per pollen grain permits a male to claim the eggs in more archegonia, preventing some fertilizations by other males and thus tending to counter SPE.

Hypotheses concerning female choice and phenomena generated by both sexes

Third-order hypotheses

(*)H_{51}. Short delays of fertilization are controlled by females; intermediate delays are controlled by males; very long delays have components of both.

*H_{52}. Females can reassert control by heterochromatinization of DNA (H_{40A}, H_{40C}) or by increasing ma-

ternal haploid complements in the endosperm (H_{40A}, H_{40B}). (Partly testable.)

Miscellaneous hypotheses

?*H_{53}. Outcrossed offspring are more likely to be specially adapted for long-distance dispersal than are inbred offspring.

*H_{54}. The occurrence of intense competition within the genome (especially that occurring between loci) favors asexual reproduction by individuals (this hypothesis not included in the theory map).

We have presented evidence that resource limitation of seed production is common; limitation by quantity of pollen appears to be relatively rare. These two conditions should generate divergent trends for several aspects of mating. Another likely possibility—that the quality of available pollen affects female reproductive patterns—must also be considered. For instance, in species not normally limited by pollen quantity the extent of acceptance of inbred zygotes may increase with an increase in the maternal resource budget or decrease with increased availability of pollen. On the other hand, if selection favors a set, possibly mixed, parentage for the seed crop, degree of inbreeding may not change at all. When resources are limiting and zygote abortion is sequential, its rate should tend to decrease during the season, because this procedure allows females to maximize production of high quality offspring and still produce a seed crop of optimal size. Changes in the rate of abortion should be affected by the abundance of high quality pollen, so the experimental manipulations that provide high pollen quality throughout the season should result in much less intraseason change in abortion rate (because the initial rate would be relatively low) than those that provide low quality (because the initial rate would be relatively high). When abortion is sequential, particular pollen types (or pollen from particular sources) should experience higher acceptance rates when they are late to fertilize than when they are early. The extent of variability in acceptance should vary among types for particular females; that is, an individual female may be uniformly accepting of some types but only accept others when they are late to fertilize. Species that delay fertilization for longer than one breeding season may be able to use information on pollen or resource availability in the second season to affect patterns of abortion of zygotes formed during the preceding one; this ability is most likely to occur in species with pollen or resource levels that vary greatly over several years.

In pollen-limited species abortion should be practiced only against zygotes with genetic incompatibilities or deficiencies. These species should display a lower tendency toward delayed fertilization and sequential abortion. They should increase selectivity when pollen levels increase, but they are less likely than resource-limited species to have well-developed capacities for zygote selectivity.

Given the fundamental importance of the concept of female choice, there is a distinct paucity of empirical studies dealing with female ability to discriminate on the basis of genetic quality or complementarity (section 3.1). Evidence for differential selectivity as a function of pollen quality and quantity might be obtained by manipulating availability of certain types of pollen or by comparing the frequencies of strategically chosen alleles in progenies of individual females and then comparing them to the frequency of available paternal alleles. Another promising approach is to compare phenotypes of offspring produced under varying pollen conditions (with other environmental variables held constant, of course) (e.g., Ter-Avanesian, 1978). Population-wide evidence for the transmission of particular alleles in apparently Mendelian proportions does not constitute good evidence for lack of mate selectivity of individual females. For example, selecting by a criterion of genetic complementarity, some females might favor a particular allele, whereas others might act to eliminate it from their progenies. The net effect may be that the allele is passed on in Mendelian proportions, apparently approximating a Hardy-Weinberg equilibrium. In turn, this might lead to the incorrect inference that females are insensitive to the occurrence of the allele or unable to respond to its presence when choosing mates. In this context the repeated occurrence of particular alleles in Mendelian proportions in populations would seem to suggest at most that those alleles may have reached some equilibrium frequency. Evidence for mate selectivity, differential abortion, and many other evolutionary and ecological processes require monitoring the activities of individuals rather than relying on inference from population patterns. (For further discussion of the inferential fallacy, see Alker, 1969; Allardt, 1969; Burley, in press.) Another limitation of relying on deviation from Hardy-Weinberg expectations to discern evidence of selection is that the very large samples required to detect relatively small deviations are seldom obtained (Sokal and Rohlf, 1969; Mulcahy and Kaplan, 1979).

We have suggested functions for various traits, some of which are nearly universal gymnosperm or angiosperm features. For example, SPE facilitates efficient female choice despite lack of prezygotic abilities to detect pollen quality. Double fertilization has the potential to reduce abortion rates. Innovations such as these seem to be so successful sometimes that they surpass all alternative and counter "strategies"; they appear to be evolutionarily stable strategies (Hamilton, 1967; Maynard Smith and Price, 1973; Maynard Smith, 1974; Parker, 1974). Once such traits become fixed, our opportunities for directly investigating the forces responsible for their evolution are weakened appreciably. We may be able to build a coherent, logical scheme that accounts for their occurrence, but we are unlikely to be able to actively investigate and test our ideas on traits that are practically invariant.

For this reason it is most important to focus on traits that are not invariant. In our examination of the literature we have sought evidence of intraspecific variability. Although we have found very little, we are confident that in many cases this results from lack of attention to, rather than lack of occurrence of, the phenomenon. Pollen size, for example, is acknowledged to be intraspecifically variable, but there is much greater attention to its mean size than to its variability, and virtually no attention to the distribution of its size in most species. Mean pollen diameter for a species or population may be valuable as a tool to species identification of pollen, but for those whose research is not strongly taxonomically oriented, this information is considerably less important. If size variation occurs, we would expect to find individual variation in the contents of pollen grains and ensuing success in avoiding abortion and perhaps in seed germinability and seedling establishment. Degrees of variation in pollen size may be constrained by other factors, including the pollen delivery system; wind-borne pollen may be less variable than zoophilous pollen, and postfertilization selection of embryos (found particularly in gymnosperms) might favor the endowment of at least some pollen grains of a male with greater investments.

That the search for typologies has superceded interest in variability at other levels of research is also apparent. We would like to know, for example, about individual variation in transmission of *B* chromosomes and about possible variability in the multiplication of maternal and paternal nuclei within the endosperm. Is there intraspecific variation in the time of onset of gene action affecting zygote development rates? We predict that events transpiring at the level of the cell will show considerable variability from species to species and at least some variation within species.

We have suggested several possible forms of interaction

among siblings, between parents and offspring, and between potential mates. The evolutionary outcome of sibling interactions depends on their degree of relatedness. Virtually no data exist about such matters, however. Can sibling pollen grains interact synergistically but interfere with unrelated pollen? When multiple seedlings germinate from a single seed, is their fate affected by whether or not they are identical? Even when embryos have identical nuclei, are there cytoplasmic differences among them that equip them for different roles and determine their fate?

Appendix

Appendix. Summary of "gymnosperm" reproductive characteristics (PT = pollen tube; arch. = archegonium or archegonia)

	SPE	1°CPE	2°CPE	Male MI
CONIFERALES				
Araucariaceae				
Agathis australis	3-25 arch., usually 8-15	–	–	♂ cytoplasm; conflicting reports on ♂ nuclei
Araucaria brasiliensis (*angustifolia*)	6-15+ arch., usually 5-8	–	–	♂ cytoplasm; sometimes ♂ nuclei
A. araucana	5-14 arch.			
Cupressaceae	archegonial complexes			
Actinostrobus pyrimidalis	25-30 arch.	variable reports		♂ cytoplasm
Callitris spp. (for the genus)	17-20 arch., sometimes 2-3 ♀ gametophytes	+	+	♂ cytoplasm; no nuclei?
Chamaecyparis nootkatensis	5-19 arch., usually 9			
C. lawsoniana	6-8 arch.			♂ cytoplasm & organelles
C. obtusa		+	+	

Delayed fertilization	Sex expression	Misc.	References
~13 mo.	monoec.		Buchholz, 1920b, 1929; Eames, 1913; Favre-Duchartre, 1966; Roy Chowdhury, 1962
~12 mo.	dioec.	sequential maturation of arch.; 3+ get fertilized	Buchholz, 1920b, 1929; Burlingame, 1913, 1914, 1915; Favre-Duchartre, 1962, 1966; Konar & Oberoi, 1969a; Roy Chowdhury, 1962
11-12 mo.	usually dioec.		Favre-Duchartre, 1962, 1970; Konar & Oberoi, 1969a
~3 mo.	monoec.	2 functional sperm per PT	Saxton, 1913a; Buchholz, 1918, 1929; Doyle & Brennan, 1972; Favre-Duchartre, 1970
12-19+ mo. depending on species	monoec.	2 functional sperm; PT induces arch.	Doyle & Brennan, 1972; Baird, 1953; Saxton, 1909, 1910b, 1913a; Looby & Doyle, 1940; Favre-Duchartre, 1966, 1974
3 mo.	monoec.		Owens & Molder, 1975a
	monoec.		Favre-Duchartre, 1966, 1970; Chesnoy, 1973, 1977
	monoec.		Buchholz, 1932b

	SPE	1°CPE	2°CPE	Male MI
C. pisifera	5-12 arch., usually 7	+		
Cupressus arizonica	3-13 arch., usually 6-8	+	–	♂ cytoplasm &/or nuclei
C. funebris	10-13 (17) arch.	+		♂ cytoplasm &/or nuclei
C. goveniana				♂ cytoplasm &/or nuclei
C. sempervirens	14-17 arch.			♂ cytoplasm &/or nuclei
Juniperus chinensis		+	+	
J. communis	4-10 arch.	+	+	♂ cytoplasm & organelles
J. oxycedrus	>10 arch.			(no ♂ nuclei)
J. squamata		+		
J. virginiana	4-14 arch.	+		♂ cytoplasm & starch; no nuclei
Libocedrus decurrens	10-15 arch.	+	+	♂ cytoplasm; no nuclei

Delayed fertilization	Sex expression	Misc.	References
	monoec.		Sugihara, 1938; Johansen, 1950; Doyle & Brennan, 1972
	monoec.	up to 14 functional sperm	Chesnoy, 1977; Doak, 1932, 1937; Sugihara, 1956
	monoec.	2 functional sperm; 4-5 PT/ ovule	Konar & Banerjee, 1963; Sugihara, 1956; Maheshwari & Sanwal, 1963; Mehra & Sircar, 1949
	monoec.	many functional sperm	Doak, 1932; Sugihara, 1956
	monoec.	up to 11 functional sperm	Mehra & Malhotra, 1947; Konar & Bannerjee, 1963; Sugihara, 1956
~11 wks.	dioec.		Tang, 1948a
12-13 mo.	usually dioec.	2 functional sperm	Favre-Duchartre, 1970; Johansen, 1950; Doyle & Brennan, 1972; Chesnoy, 1969; Tang, 1948a; Ottley, 1909; Cook, 1939
	dioec.		Favre-Duchartre, 1966, 1970
			Doyle & Brennan, 1972
variable reports: 1-2½ mo.	usually dioec.	2 functional sperm	VanHaverbeke & Read, 1976; 1939; Ottley, 1909; Favre-Duchartre, 1966
~2 mo.	usually monoec.		Buchholz, 1929; Chesnoy, 1977; Lawson, 1907b; Konar & Oberoi, 1969a; Favre-Duchartre, 1966

	SPE	1°CPE	2°CPE	Male MI
Tetraclinis articulata	usually 9-10 arch.	+		♂ cytoplasm
Thuja occidentalis	5-7 arch., usually 6	+ (−)	variable reports	variable ♂ nuclei; a little ♂ cytoplasm
T. (Biota) orientalis	15-28 arch., usually 23-24	+		♂ cytoplasm & organelles
T. plicata	7-9 arch.	−	−	♂ nuclei
Thujopsis dolabrata	sometimes 2 ♀ gametophytes, 5-12 arch., usually 7-9	+ (rare)	+	♂ cytoplasm
Widdringtonia cupressoides	up to 100+ arch., usually 40-70	+		no ♂ nuclei
W. juniperoides	~200 arch.	+		
Pinaceae				
Abies amabilis	2-3 arch.	+ (rare)	−	
A. balsamea	1-4 arch.	10%	rare	♂ nuclei

Delayed fertilization	Sex expression	Misc.	References
~3 mos.	monoec.	2 functional sperm	Chesnoy, 1977; Saxton, 1913b; Konar & Oberoi, 1969a
	monoec.		Land, 1902; Martin, 1950; Buchholz, 1920b; Maheshwari & Sanwal, 1963
~2 mo.	monoec.		Chesnoy, 1977; Doyle & Brennan, 1972; Lawson, 1907b; Singh & Oberoi, 1962; Martin, 1950; Chesnoy & Thomas, 1971; Favre-Duchartre, 1966; Mehra & Sircar, 1949
~2½ mo.	monoec.	several PT reach arch.	Owens and Molder, 1980b
~2 mo.	?		Doyle & Brennan, 1972; Sugihara, 1939; Johansen, 1950; Maheshwari & Sanwal, 1963
14-15 mo.	usually monoec.	2 functional sperm; PT induces arch.	Favre-Duchartre, 1966; Saxton, 1909, 1910a; Moseley, 1943
	usually monoec.		Saxton, 1934b; Doyle & Brennan, 1972
~2 mo.	monoec.		Owens & Molder, 1977b; Hutchinson, 1924
	monoec.		Hutchinson, 1924; Favre-Duchartre, 1966, 1970; Buchholz, 1920a, 1930

	SPE	1°CPE	2°CPE	Male MI
A. firma		28%	occasional	
A. grandis		rare	–	
A. koreana				
A. nordmanniana	2-4 arch.			
A. venusta			+	
Cedrus atlantica	4-6 arch.			
C. deodora	3-5 arch.	+	+	♂ nuclei & probably cytoplasm
C. libani		+	+	
Keteleeria davidiana	5-10 arch., usually 7	+	rare	variable nos. ♂ nuclei
K. evelyniana	7-12 arch., usually 9-11			
Larix decidua	1-5 arch.	+	–	some ♂ mitochondria; all ♂ plastids
L. europea	about 5 arch.	–	–	
L. leptolepis	5 arch.			
L. occidentalis	2-5 arch.	–	–	♂ cytoplasm

Delayed fertilization	Sex expression	Misc.	References
	monoec.		Sugihara, 1947c
	monoec.		Hutchinson, 1924
~2 mo.	monoec.		Doyle & Kane, 1942
~2 mo.	monoec.		Doyle & Kane, 1942; Favre-Duchartre, 1970
	monoec.		Buchholz, 1942
~9 mo.	monoec.		Smith, 1923
~8 mo.	usually dioec.		Favre-Duchartre, 1966; Roy Chowdhury, 1961
	monoec.		Buchholz, 1920a, 1930
	monoec.		Favre-Duchartre, 1966; Sugihara, 1943b
~3½ mo.	monoec.		Wang, 1948b
6-8 wks.	monoec.		Camefort, 1967, 1968, 1969; Chesnoy, 1977; Chesnoy & Thomas, 1971; Schopf, 1943; Christiansen, 1969b; Barner & Christiansen, 1960
	monoec.		Buchholz, 1920a; Favre-Duchartre, 1970
6-8 wks.	monoec.		Christiansen, 1972b; Schopf, 1943
6-8 wks.	monoec.		Owens and Molder, 1979b

	SPE	1°CPE	2°CPE	Male MI
Picea abies	1-7 arch., average 3	–	–	
P. excelsa	2-7 arch., usually 4	–	–	♂ nuclei & cytoplasm
P. glauca	1-4 arch., usually 3	–	–	
P. mariana		–	–	
P. pungens				
P. sitchensis	1-3 arch., usually 2	–	–	♂ nuclei
Pinus spp. (for the genus)		+	+	♂ cytoplasm & nuclei
Pinus banksiana	2-3 arch.	+	+	
P. gerardiana	1-4 arch.	+	+	
P. lambertiana	3-5 arch., usually 5	+	+	+
P. laricio	2-6 arch.	+	+	no ♂ organelles
P. monophylla	3-5 arch., usually 3	+	+	+
P. nigra				♂ cytoplasm & organelles

Delayed fertilization	Sex expression	Misc.	References
26-36 d.	monoec.	sequential fertilization of arch.; 2 sperm; parthenocarpic	Sarvas, 1968; Mikkola, 1969
6 wks.	monoec.	sequential fertilization of arch.	Buchholz, 1920a, 1930; Lill, 1976; Miyake, 1903; Doyle & Brennan, 1972
3 wks.	monoec.		Owens and Molder, 1979a
	monoec.	sequential fertilization of arch.	Buchholz, 1920a
4-5 wks.			Hanover, 1975
7 wks.	monoec.		Owens and Molder, 1980a
12+ mo.	monoec.		Favre-Duchartre, 1966; Allen, 1946; Lill, 1976; Buchholz, 1918; Maheshwari & Konar, 1971; but compare Vazart, 1958
	monoec.	some twinning	Buchholz, 1918; Clare & Johnstone, 1931; Johnstone, 1940
12+ mo.	monoec.		Konar, 1962; Lill, 1976
12+ mo.	monoec.		Haupt, 1941; Lill, 1976
	monoec.		Camefort, 1966a,b; Lill, 1976
12+ mo.	monoec.		Haupt, 1941; Lill, 1976
12+ mo.	monoec.		Chesnoy, 1977; Chesnoy & Thomas, 1971; McWilliam & Mergen, 1958

	SPE	1°CPE	2°CPE	Male MI
P. palustris		+		
P. pinaster		+		no ♂ organelles
P. radiata	1-4 arch., often 1	+	+	
P. resinosa	1-5 arch.	+		+
P. rigida	1-5 arch.	+		+
P. strobus	1-5 arch., usually 2-3	+		+
P. sylvestris	1-4 arch.	+		+
P. wallichiana	1-2 arch.	+		+
Pseudolarix amabilis (*kaempferi*)	4-7 arch.	variable reports	variable reports	
Pseudotsuga douglasii	4-6 arch.			♂ cytoplasm
P. menziesii		–	–	♂ cytoplasm & nuclei
P. taxifolia	5-8 arch.	–	–	♂ nuclei & cytoplasm
Tsuga canadensis	usually 2-4 arch.	+	+	♂ nuclei & a little cytoplasm
T. caroliniana			–	

Delayed fertilization	Sex expression	Misc.	References
~15 mo.	monoec.		Snyder *et al.*, 1977
	monoec.		Camefort, 1966a
~15 mo.	monoec.		Lill, 1976; Johansen, 1950; Burdon & Zabkiewicz, 1973
	monoec.		Ferguson, 1904; Lill, 1976
	monoec.		Ferguson, 1904; Lill, 1976
	monoec.		Ferguson, 1901, 1904, 1913; Lill, 1976
	monoec.		Owens & Molder, 1977a; Dixon, 1894; Blackman, 1898; Lill, 1976
	monoec.		Konar & Ramchandani, 1958; Lill, 1976
variable reports: 5-10 wks.	monoec.		Buchholz, 1930; Mergen, 1976; Favre-Duchartre, 1970; Miyake & Yasui, 1911; Hall & Brown, 1977
~2 mo.	monoec.		Lawson, 1909
variable reports: 6-10 wks.		2 functional sperm	Allen & Owens, 1972; Christiansen, 1969a,b, 1972b
9-10 wks.			Buchholz, 1920a; Allen, 1946
~6 wks.	monoec.		Sterling, 1948b; Buchholz, 1920a, 1930; Murrill, 1900; Favre-Duchartre, 1966
	monoec.		Buchholz, 1930

	SPE	1°CPE	2°CPE	Male MI
T. heterophylla	2-5 arch., usually 3	+	–	+
T. mertensiana	usually 2-4 arch.	+	+	
Podocarpaceae (for the family)	often 2 ♀ gametophytes			♂ nuclei
Dacrydium bidwelli	2 arch., rarely 1-3 (very constant)	+	+	♂ cytoplasm & nuclei
D. colensoi	1-4 arch.	–		
D. cupressinum	3 arch.	+		
D. laxifolium	usually 2 arch., rarely 3-4	–	–	
Microcachrys tetragona	5-6 arch.	–	–	
Pherosphaera fitzgeraldi	3-4 arch.	+	+	
P. hookeriana	2-6 arch., usually 3	+	+	
Phyllocladus alpinus	1-4 arch., usually 2	–	–	♂ cytoplasm & nuclei
P. glaucus	2 arch.	–	–	

Delayed fertilization	Sex expression	Misc.	References
~6 wks.	monoec.		Buchholz, 1930; Stanlake & Owens, 1974; Doyle & O'Leary, 1935b
~6 wks.	monoec.		Owens & Molder, 1975b; Buchholz, 1930
			Sinnott, 1913; Doyle & Brennan, 1971; Allen, 1946; Chamberlain, 1935; Boyle & Doyle, 1953
~12 mo.	dioec.		Sinnott, 1913; Quinn, 1966a
~13 mo.?	monoec. or dioec.		Quinn, 1966b; Maheshwari & Sanwal, 1963
			Doyle, 1954; Sinnott, 1913
~12 mo.	monoec. or dioec.		Quinn, 1965
	monoec.	2 functional sperm	Boyle & Doyle, 1953; Lawson, 1923b
	monoec.		Lawson, 1923a,b
	monoec.		Doyle, 1954; Elliott, 1948, 1950; Lawson, 1923a
	mostly dioec.		Stiles, 1912; Buchholz, 1929, 1941b; Boyle & Doyle, 1953; Sinnott, 1913; Kildahl, 1908; Young, 1910; Holloway, 193
	dioec.		Holloway, 1937; Stiles, 1912; Maheshwari & Sanwal, 1963

	SPE	1°CPE	2°CPE	Male MI
Saxegothea conspicua	2-3 arch., often 2 ♀ gametophytes	~15%? (rare)		♂ cytoplasm & nuclei
Podocarpus (listed by sections, as in Doyle & Brennan)				
(Stachycarpus)				
P. andinus	usually 2, sometimes 3 arch.	–		♂ cytoplasm & nuclei
P. spicatus	usually 2, sometimes 3 arch.	–	–	
P. ferrugineus	usually 2, sometimes 3 arch.	–	–	
(Afrocarpus)				
P. usumbarensis	usually 4-5 arch.	–	–	
P. falcatus	7-10 arch.	rare	–	
P. gracilior	2-22 arch., usually 10-22, sometimes in complexes	+		
(Sundacarpus)				
P. amarus		+ (some)	+ (occasional)	
(Nageia)				
P. nagi	up to 19 arch.	+	+	

Delayed fertilization	Sex expression	Misc.	References
~12+ mo.	usually monoec.		Buchholz, 1941b; Looby & Doyle, 1939; Doyle & O'Leary, 1935a; Boyle & Doyle, 1953; Stiles, 1912; Wardlaw, 1955; Johansen, 1950; Chesnoy, 1977; Favre-Duchartre, 1966
>12 mo.	usually dioec.	1 functional sperm	Boyle & Doyle, 1953; Looby & Doyle, 1944a,b; Doyle, 1954; Favre-Duchartre, 1966
~15 mo.	dioec.		Boyle & Doyle, 1953; Buchholz, 1936, 1941b, 1920b; Doyle, 1954; Sinnott, 1913
~15 mo.	dioec.		Boyle & Doyle, 1953; Buchholz, 1936, 1941b, 1920b; Doyle, 1954; Sinnott, 1913
~4 mo.	dioec.		Buchholz, 1936
no pollen "resting period"		PT induces arch.	Osborn, 1960; Konar & Oberoi, 1969b; Favre-Duchartre, 1970
~2½ mo., but no pollen "resting period"		PT induces arch.; 2 sperm; as many as 5 PT reach ovule	Doyle & Brennan, 1971; Konar & Oberoi, 1969b
			Doyle, 1954; Buchholz, 1941b
5 wks.			Stiles, 1912; Boyle & Doyle, 1953; Buchholz, 1941b; Wang, 1950

	SPE	1°CPE	2°CPE	Male MI
(Dacrycarpus)				
P. dacrydioides	3-12 arch., usually 5-7	+	–	
P. imbricatus		variable reports	+	
(Eupodocarpus)				
P. acutifolius	usually 2 arch., up to 14	relatively rare	very rare	
P. glomeratus		+	+?	
P. halli	3-6 arch., usually 2 are fertilized	relatively rare	very rare	♂ cytoplasm
P. henkelii	usually 1 arch., rarely 4 or 5	+	+	
P. macrophyllus	5-14 arch.	variable	variable	
P. nivalis	3-6 arch., usually 2 are fertilized	relatively rare	very rare	♂ cytoplasm & nuclei
P. purdeanus		variable reports	+	

Delayed fertilization	Sex expression	Misc.	References
variable reports: 5-8 wks.			Boyle & Doyle, 1953; Buchholz, 1941b; Doyle, 1954; Sinnott, 1913; Quinn, 1964
			Buchholz, 1941b; Doyle, 1954
~1 mo.			Sinnott, 1913; Brownlie, 1953
			Buchholz, 1941b; Johansen, 1950
~1 mo.			Brownlie, 1953; Sinnott, 1913
			Coertze *et al.*, 1971
		1 functional sperm	Buchholz, 1941b; Coker, 1902; Stiles, 1912; Boyle & Doyle, 1953; Wardlaw, 1955; Johansen, 1950
~1 mo.			Favre-Duchartre, 1966; Sinnott, 1913; Brownlie, 1953; Buchholz, 1920b; Doyle, 1954; Boyle & Doyle, 1953
			Doyle & Brennan, 1971; Buchholz, 1941b

	SPE	1°CPE	2°CPE	Male MI
P. totara	3-6 arch., usually 2 are fertilized	relatively rare	very rare	♂ cytoplasm
P. urbanii		+	+	
Taxaceae				
Amentotaxus argotaenia	2-3 arch.	–?	–?	
Austrotaxus spicata	3-5 arch., often 2 ♀ gametophytes	–	–	
Pseudotaxus chienii	up to 6 ♀ gametophytes, 3-7 arch.			♂ cytoplasm & nuclei
Taxus baccata	sometimes >1 ♀ gametophyte, 5-11 arch. in complexes	–	rare	variable reports
T. canadensis	sometimes >1 ♀ gametophyte, 4-8 arch.	–		♂ cytoplasm & nuclei
T. cuspidata	6-25+ arch., usually 8-14, in complexes	–?	occasional	♂ nuclei at least
T. wallichiana		–		

Delayed fertilization	Sex expression	Misc.	References
4-6 wks.			Brownlie, 1953; Sinnott, 1913; Buchholz, 1941b (Johansen [1950] says fertilization occurs after seed is shed).
			Buchholz, 1941b; Wardlaw, 1955
	dioec.		Doyle & Brennan, 1971; Favre-Duchartre, 1970
	?		Buchholz, 1940a; Saxton, 1934a, 1936
5 wks.	dioec.		Chen & Wang, 1979
2½-3 mo.	dioec.	1 functional sperm	Allen & Owens, 1972; Buchholz, 1929; Saxton, 1936; Favre-Duchartre, 1962, 1966; Sterling, 1948a; Dupler, 1917; Chesnoy, 1977; Doyle & Brennan, 1971; Stanlake & Owens, 1974; Vazart, 1958
~1 mo., but variable	dioec.		Allen, 1946; Doyle & Brennan, 1971; Dupler, 1917
	dioec.		Doyle & Brennan, 1971; Sterling, 1940a
	dioec.	sequential maturation of arch.	Favre-Duchartre, 1974; Doyle & Brennan, 1971

	SPE	1°CPE	2°CPE	Male M1
Torreya californica (*myristica*)	2-6 arch., usually 3	+	+	variable reports
T. grandis	1-3 arch., usually 1	+	+	
T. nucifera	2-5 arch., usually 3	+	+	variable reports
T. taxifolia	1 arch.	+	+	variable reports
"Cephalotaxaceae"				
Cephalotaxus drupacea	3-7 arch., usually 4	+ (rare?)	+	♂ cytoplasm, nuclei, & organelles
C. fortunei	2-5 arch., usually 3	–	+	♂ cytoplasm & nuclei
Taxodiaceae (see review in Doyle & Brennan 1971)				
Arthrotaxis selaginoides	16-36 arch. in complex	–	–	♂ cytoplasm; no nuclei
Cryptomeria japonica	8-37 arch. in complex	+	+	♂ mitochondria, plastids, & starch

Delayed fertilization	Sex expression	Misc.	References
3-4 mo.	dioec.	1 functional sperm	Allen & Owens, 1972; Buchholz, 1940a; Robertson, 1904
3-4 wks.	dioec.	1 functional sperm	Tang, 1948b
	dioec.	1 functional sperm	Buchholz, 1940a; Tahara, 1940
4½ mo.	dioec.	1 functional sperm	Buchholz, 1929, 1940a; Coulter & Land, 1905; Stanlake & Owens, 1974; Maheshwari & Sanwal, 1963
14-15 mo.	dioec.	1 functional sperm; sequential maturation of arch.	Allen & Owens, 1972; Singh, 1961; Gianordoli, 1974; Favre-Duchartre, 1957, 1962, 1966, 1974; Lawson, 1907a; Vazart, 1958
14 mo.	dioec.	1 functional sperm; sequential maturation of arch.; 1 PT/arch.	Buchholz, 1925, 1920b, 1930, 1933; Coker, 1907
	monoec.	PT induces arch.	Favre-Duchartre, 1966; Chesnoy, 1977; Brennan & Doyle, 1956; Saxton & Doyle, 1929; Maheshwari & Sanwal, 1963
~3-3½ mo.	monoec.	>1 functional sperm; parthenocarpy	Singh & Chatterjee, 1963; Hashizume, 1973b; Johansen, 1950; Sugihara, 1969; 1947a; Dogra, 1966b; Buchholz, 1932a; Lawson, 1904b; Chesnoy, 1977; Vazart, 1958

	SPE	1°CPE	2°CPE	Male MI
Cunninghamia lanceolata	10-16 arch. in complex	+	+	♂ cytoplasm & starch
C. konishii		+	+?	♂ cytoplasm & starch
Glyptostrobus pensilis		+	–	
Metasequoia glyptostroboides	8-10 arch., in complex	+	–	♂ nuclei (variable)
Sequoia sempervirens	>1 ♀ gametophyte, many arch. in complex	+	rare	variable reports
Sequoiadendron giganteum	many arch.	+	+	some ♂ cytoplasm (variable); no nuclei
Taiwania cryptomeroides	4-9 arch. in complex	+		
Taxodium distichum	6-34+ arch., usually 10-20	+	–	♂ cytoplasm & starch
T. mucronatum	12-20 arch., usually 16	+	+	♂ cytoplasm & starch
- - - - - - -				
Sciadopitys verticillata	4-8 arch. usually 4, not in complex	+	variable	♂ cytoplasm & variable nuclei

Delayed fertilization	Sex expression	Misc.	References
~5 mo.	monoec.		Sugihara, 1943a, 1947b; Dogra, 1966b; Johansen, 1950; Buchholz, 1940b; Miyake, 1908; Chesnoy, 1977; Vazart, 1958
	monoec.		Sugihara, 1947b
	monoec.		Wang, 1948a
	monoec.	sticky, clumped pollen	Wang & Chien, 1964; Johnson, 1974; Favre-Duchartre, 1966
variable reports; 4-7 mo.	monoec.	2 functional sperm; PT induces arch.	Looby & Doyle, 1937, 1942; Lawson, 1904a; Buchholz, 1939b; Shaw, 1896; Favre-Duchartre, 1966
~6 mo.	monoec.		Looby & Doyle, 1937, 1942; Buchholz, 1939a; Favre-Duchartre, 1966
	monoec.		Sugihara, 1941
~4 mo.	monoec.	2 functional sperm	Coker, 1903; Kaeiser, 1940
~2 mo.	monoec.		Brennan & Doyle, 1956; Vasil & Sahni, 1964; Dogra, 1966b; Chesnoy, 1977
~14 mo.	monoec.	arch. mature before PT begins to grow	Gianordoli, 1964; Buchholz, 1931, 1920b; Lawson, 1910; Johansen, 1950; Favre-Duchartre, 1966; Maheshwari & Sanwal, 1963

	SPE	1°CPE	2°CPE	Male MI
CYCADALES				
Bowenia spectabilis	~6 arch.			♂ cytoplasm
Cycas circinalis	1-7 arch.			
C. rumphii	3-6 arch.			
Dioon edule	up to 10 arch.			♂ cytoplasm
Macrozamia reidlei	3-15 arch., usually 4-7	–	+	♂ cytoplasm
M. spiralis	3-8 arch., usually 4-5	+	+?	♂ cytoplasm
Microcycas calocoma	hundreds of arch.			
GINKGOALES				
Ginkgo biloba	2-5 arch., usually 2	occasional		♂ cytoplasm; no nuclei
GNETALES				
Gnetum funiculare				

Delayed fertilization	Sex expression	Misc.	References
~4½ mo.	dioec.	1 functional sperm	Lawson, 1926
	dioec.		Rao, 1961
~13 mo.	dioec.	polyspermy	deSilva & Tambiah, 1952
			Chamberlain, 1910; Favre-Duchartre, 1970
~4 mo.	dioec.		Baird, 1939
⩾ 2 mo.	dioec.	polyspermy	Brough & Taylor, 1940
	dioec.	16-20 functional sperm	Reynolds, 1924; Downie, 1928; Caldwell, 1907
5 mo.	dioec.	fertilization may occur after detachment of megasporangium	Favre-Duchartre, 1956, 1958, 1962, 1966; Lyon, 1904; Cook, 1903; Lee, 1955; Maheshwari & Sanwal, 1963; Vazart, 1958; Faegri & van der Pijl, 1979; Eames, 1955
	dioec.	2 functional sperm	Haining, 1920; Favre-Duchartre, 1970

	SPE	1°CPE	2°CPE	Male MI
G. gnemon	1-3 eggs, usually 2	+	+	
G. ula	often 2 ♀ gametophytes, 1-6 eggs	+	+	♂ nuclei at least
G. africanum	sometimes >1 embryo sac			
EPHEDRALES				
Ephedra distachya	1-11 arch., usually 5-6	+		♂ cytoplasm & nuclei
E. foliata	+	+		♂ cytoplasm & nuclei
E. intermedia				
E. trifurca		+		
WELWITSCHIALES				
Welwitschia mirabilis	+			

NOTE: Data on sex expression from Dallimore and Jackson (1948) and Favre-Duchartre (1970).

Delayed fertilization	Sex expression	Misc.	References
	dioec.	>1 functional sperm; up to 12 zygotes/ gametophyte	Bower, 1882; Sanwal, 1962; Coulter, 1908; Favre-Duchartre, 1970
1-2 wks.	dioec.	some polyploid "endosperm"	Swamy, 1973; Vasil, 1959; Favre-Duchartre, 1970
		2 functional sperm, no double fertilization	Waterkeyn, 1954
~1 mo.	usually dioec.	1 functional sperm	Chamberlain, 1935; Berridge & Sanday, 1907; Favre-Duchartre, 1962, 1966, 1970
	usually dioec.		Khan, 1943
	sometimes monoec.		Mehra, 1950
<1 day	usually dioec.		Land, 1907
	dioec.		Pearson, 1910; Johansen, 1950

Literature Cited

Aalders, L. E., I. V. Hall, and F. R. Forsyth. 1969. Effects of partial defoliation and light intensity on fruit-set and berry development in the lowbush blueberry. *Hort. Res.* 9:124-129.

Addicott, F. T., and R. S. Lynch. 1955. Physiology of abscission. *Annu. Rev. Plant Physiol.* 6:211-238.

Aker, C. L., and D. Udovic. 1981. Oviposition and pollination behavior of the yucca moth, *Tegeticula maculata* (Lepidoptera: Prodoxidae), and its relation to the reproductive biology of *Yucca whipplei* (Agavaceae). *Oecologia* 49:96-101.

Alexander, R. D. 1974. The evolution of social behavior. *Annu. Rev. Ecol. Syst.* 5:325-383.

Alexander, R. D. 1979. *Darwinism and Human Affairs.* Seattle: Univ. of Washington Press.

Alexander, R. D., and G. Borgia. 1978. Group selection, altruism, and the levels of organization of life. *Annu. Rev. Ecol. Syst.* 9:449-474.

Alexander, R. D., and G. Borgia. 1979. On the origin and basis of the male-female phenomenon. Pp. 412-440 in M. S. Blum and N. A. Blum, eds., *Sexual Selection and Reproductive Competition in Insects.* New York: Academic Press.

Alexander, R. D., and P. W. Sherman. 1977. Local mate competition and parental investment in social insects. *Science* 196:494-500.

Alker, H. R., Jr. 1969. A typology of ecological fallacies. Pp. 69-86 in M. Dogan and S. Rokkan, eds., *Quantitative Ecological Analysis in the Social Sciences.* Cambridge, Mass.: MIT Press.

Allard, R. W., S. K. Jain, and P. L. Workman. 1968. The

genetics of inbreeding populations. *Adv. Genet.* 14:55-131.

Allardt, E. 1969. Aggregate analysis: the problems of its informative value. Pp. 53-68 in M. Dogan and S. Rokkan, eds., *Quantitative Ecological Analysis in the Social Sciences.* Cambridge, Mass.: MIT Press.

Allen, G. S. 1946. Embryogeny and development of the apical meristems of *Pseudotsuga.* I. Fertilization and early embryogeny. *Amer. J. Bot.* 33:666-676.

Allen, G. S., and J. N. Owens. 1972. *The Life History of Douglas Fir.* Ottawa: Information Canada.

Allen, P. H., and K. B. Trousdell. 1961. Loblolly pine seed production in the Virginia-North Carolina coastal plain. *J. Forestry* 59:187-190.

Allen, R. M. 1953. Release and fertilization stimulate longleaf pine cone crop. *J. Forestry* 51:827.

Anderson, L. E. 1936. Mitochondria in the life cycles of certain higher plants. *Amer. J. Bot.* 23:490-500.

Andersson, E. 1965. Cone and seed studies in Norway spruce [*Picea abies* [L.] Kurst). *Studia Forest. Suecica* 23:1-214.

Andersson, E., R. Jansson, and D. Lindgren. 1974. Some results from second generation crossings involving inbreeding in Norway spruce (*Picea abies*). *Silv. Genet.* 23:34-43.

Antonovics, J. 1968. Evolution in closely adjacent plant populations. V. Evolution of self-fertility. *Heredity* 23:219-238.

Antonovics, J. 1976. The nature of limits to natural selection. *Ann. Missouri Bot. Garden* 63:224-247.

Arasu, N. P. 1968. Self-incompatibility in angiosperms. *Genetica* 39:1-24.

Askew, R. R. 1971. *Parasitic Insects.* New York: American Elsevier.

Atkinson, L. R. 1938. Cytology. Pp. 196-232 in F. Verdoorn, ed., *Manual of Pteridology.* The Hague: Martinus Nijhoff.

Augspurger, C. K. 1980. Mass-flowering of a tropical shrub (*Hybanthus prunifolius*): influence on pollinator attraction and movement. *Evolution* 34:475-488.

Augspurger, C. K. 1981a. Flowering synchrony of neotropical plants. In W. D'Arcy, ed., *Natural History Studies in Panama and Central America.* Missouri Botanical Garden.

Augspurger, C. K. 1981b. Reproductive synchrony of a tropical shrub: experimental studies on effects of pollinators and seed predators on *Hybanthus prunifolius* (Violaceae). *Ecology* 62:775-788.

Austin, C. R. 1965. *Fertilization.* Englewood Cliffs, N.J.: Prentice Hall.

Austin, C. R. 1968. *Ultrastructure of Fertilization.* New York: Holt, Rinehart, Winston.

Ayonoadu, U. W., and H. Rees. 1968. The regulation of mitosis by B-chromosomes in rye. *Exper. Cell Res.* 52:284-290.

Bacchi, O. 1943. Cytological observations in citrus: III. Megasporogenesis, fertilization, and polyembryony. *Bot. Gaz.* 105:221-225.

Baetcke, K. P., A. H. Sparrow, C. H. Nauman, and S. S. Schwemmer. 1967. The relationship of DNA content to nuclear and chromosome volumes and to radio sensitivity (LD_{50}). *Proc. Nat. Acad. Sci.* (USA) 58:533-540.

Baird, A. M. 1939. A contribution to the life history of *Macrozamia reidlei. J. Roy. Soc. W. Aust.* 25:153-175.

Baird, A. M. 1953. The life history of *Callitris. Phytomorphology* 3:258-284.

Baker, H. G. 1955. Self-compatibility and establishment after "long-distance" dispersal. *Evolution* 9:347-349.

Baker, H. G. 1959. Reproductive methods as factors in speciation in flowering plants. *Cold Spring Harbor Symp. Quant. Biol.* 24:177-190.

Baker, H. G. 1963. Evolutionary mechanisms in pollination biology. *Science* 139:877-883.

Baker, H. G. 1975. Sporophyte-gametophyte interactions in *Linum* and other genera with heteromorphic incompatibility. Pp. 191-199 in D. L. Mulcahy, ed., *Gamete Competition in Plants and Animals.* Amsterdam: North-Holland.

Baker, H. G., and I. Baker. 1979. Starch in angiosperm pollen grains and its evolutionary significance. *Amer. J. Bot.* 66:591-600.

Bannister, M. H. 1965. Variation in the breeding system of *Pinus radiata.* Pp. 353-372 in H. G. Baker and G. L. Stebbins, eds., *The Genetics of Colonizing Species.* New York: Academic Press.

Banyard, B. J., and S. H. James. 1979. Biosystematic studies in the *Stylidium crassifolium* species complex. *Aust. J. Bot.* 27:27-37.

Barlow, P. W. 1972. Differential cell division in human X chromosome mosaics. *Humangenetik* 14:122-127.

Barlow, P. W., and C. G. Vosa. 1970. The effect of supernumerary chromosomes on meiosis in *Puschkinia libanotica* (Liliaceae). *Chromosoma* 30:344-355.

Barner, H., and H. Christiansen. 1960. The formation of pollen, the pollination mechanism, and the determination of the most favourable time for controlled pollinations in *Larix. Silv. Genet.* 9:1-11.

Barner, H., and H. Christiansen. 1962. The formation of pollen, the pollination mechanism, and the determination of the most favourable time for controlled pollination in *Pseudotsuga menziesii. Silv. Genet.* 11:89-102.

Barnes, B. V. 1964. Self- and cross-pollination of western white pine: a comparison of height growth of progeny. *USDA For. Serv. Res. Not.* INT-22.

Barnes, B. V., R. T. Bingham, and A. E. Squillace. 1962. Selective fertilization in *Pinus monticola* Dougl. II. Results of additional tests. *Silv. Genet.* 11:103-111.

Barnes, D. K., and R. W. Cleveland. 1963. Genetic evidence

for nonrandom fertilization in alfalfa as influenced by differential pollen tube growth. *Crop. Sci.* 3:295-297.

Bassett, I. J., and C. W. Crompton. 1968. Pollen morphology and chromosome numbers of the family Plantaginaceae in North America. *Canad. J. Bot.* 46:349-361.

Bastock, M. 1956. A gene mutation which changes a behavior pattern. *Evolution* 10:421-439.

Bateman, A. J. 1947a. Contamination of seed crops. I. Insect pollination. *J. Genetics* 48:257-275.

Bateman, A. J. 1947b. Contamination of seed crops. II. Wind pollination. *Heredity* 1:235-246.

Bateman, A. J. 1947c. Contamination in seed crops. III. Relation with isolation distance. *Heredity* 1:303-336.

Bateman, A. J. 1948. Intrasexual selection in *Drosophila*. *Heredity* 2:349-368.

Bateman, A. J. 1950. Is gene dispersion normal? *Heredity* 4:353-363.

Bateman, A. J. 1952a. Self-incompatibility systems in angiosperms. I. Theory. *Heredity* 6:285-310.

Bateman, A. J. 1952b. Variation within French bean varieties. *Ann. Appl. Biol.* 39:129-138.

Bateson, P. 1978. Sexual imprinting and optimal outbreeding. *Nature* 273:659-660.

Battaglia, E. 1964. Cytogenetics of B-chromosomes. *Caryologia* 17:245-299.

Bawa, K. S. 1980. Evolution of dioecy in flowering plants. *Annu. Rev. Ecol. Syst.* 11:15-39.

Bazzaz, F. A., R. W. Carlson, and J. L. Harper. 1979. Contribution to reproductive effort by photosynthesis of flowers and fruit. *Nature* 279:554-555.

Beale, G., and J. Knowles. 1978. *Extranuclear Genetics*. London: Edward Arnold.

Beattie, A. J. 1971. Itinerant pollinators in a forest. *Madroño* 21:120-124.

Bell, C. R. 1959. Mineral nutrition and flower to flower pollen size variation. *Amer. J. Bot.* 46:621-624.

Bell, P. R. 1975. Physical interactions of nucleus and cytoplasm in plant cells. *Endeavor* 34:19-22.

Bell, P. R. 1979. The contribution of the ferns to an understanding of the life cycles of vascular plants. Pp. 58-85 in A. F. Dyer, ed., *The Experimental Biology of Ferns.* New York: Academic Press.

Bell, P. R., and J. G. Duckett. 1976. Gametogenesis and fertilization in *Pteridium. Bot. J. Linn. Soc.* 73:47-78.

Bengtsson, B. O. 1978. Avoiding inbreeding: at what cost? *J. Theor. Biol.* 73:439-444.

Bennett, M. D. 1972. Nuclear DNA and minimum generation time in herbaceous plants. *Proc. Roy. Soc. Lond. B* 181:109-135.

Bennett, M. D. and J. B. Smith. 1972. The efforts of polyploidy on meiotic duration and pollen development in cereal anthers. *Proc. Roy. Soc. Lond. B* 181:81-107.

Benseler, R. W. 1975. Floral biology of California buckeye. *Madroño* 23:41-52.

Bentley, S., J. B. Whittaker, and A.J.C. Malloch. 1980. Field experiments on the effects of grazing by a chrysomelid beetle (*Gastrophysa viridula*) on seed production and quality in *Rumex obtusifolius* and *Rumex crispus*. *J. Ecology* 68:671-674.

Berridge, E. M., and E. Sanday. 1907. Oogenesis and embryogeny in *Ephedra distachya. New Phytol.* 6:127-134, 167-174.

Bertin, R. I. 1982a. The ecology of sex expression in red buckeye. *Ecology* 63:445-456.

Bertin, R. I. 1982b. Floral biology, hummingbird pollination, and fruit production of trumpet creeper (*Campsis radicans,* Bignoniaceae). *Amer. J. Bot.* 69:122-134.

Bertin, R. I. 1982c. Influence of father identity on fruit production in trumpet creeper (*Campsis radicans*). *Amer. Natur.* 119:694-709.

Bierzychudek, P. 1981. Pollinator limitation of plant reproductive effort. *Amer. Natur.* 117:838-842.

Bierzychudek, P. 1982. Jack and Jill in the pulpit. *Natural History* 91(3):22-27.

Bingham, R. T. 1973. Possibilities for improvement of western white pine by inbreeding. *USDA For. Serv. Res. Pap.* INT-144.

Bingham, R. T., R. J. Hoff, and R. J. Steinhoff. 1972. Genetics of western white pine. *USDA For. Serv. Res. Pap.* WO-12.

Birky, C. W. 1975a. Zygote heterogeneity and uniparental inheritance of mitochondrial genes in yeast. *Mol. Gen. Genet.* 141:41-58.

Birky, C. W. 1975b. Mitochondrial genetics in fungi and ciliates. Pp. 182-224 in C. W. Birky, P. S. Perlman, and T. J. Byers, eds., *Genetics and Biogenesis of Mitochondria and Chloroplasts.* Columbus: Ohio St. Univ. Press.

Birky, C. W. 1976. The inheritance of genes in mitochondria and chloroplasts. *Bioscience* 26:26-33.

Birky, C. W. 1978. Transmission genetics of mitochondria and chloroplasts. *Annu. Rev. Genet.* 12:471-512.

Birky, C. W., C. A. Demko, P. S. Perlman, and R. Strausberg. 1978. Uniparental inheritance of mitochondrial genes in yeast: dependence on input bias of mitochondrial DNA, and preliminary investigations of the mechanism. *Genetics* 89:615-651.

Blackman, V. H. 1898. On the cytological features of fertilization and related phenomena in *Pinus sylvestris. Phil. Trans. Roy. Soc. Lond. B* 190:395-426.

Blinkenberg, C., H. Brix, M. Schaffalitzky de Muckadell, and H. Vedel. 1958. Controlled pollination in *Fagus. Silv. Genet.* 7:116-122.

Boag, P. T., and P. R. Grant. 1981. Intense natural selection in a population of Darwin's finches (Geospizinae) in the Galapagos. *Science* 214:82-85.

Bold, H. C., C. J. Alexopoulos, and T. Delevoryas. 1980.

Morphology of Plants and Fungi. New York: Harper & Row.

Bookman, S. S. Ms. A demonstration of sexual selection in plants (*Asclepias speciosa* Torr.).

Borgia, G. 1979. Sexual selection and the evolution of mating systems. Pp. 19-80 in M. S. Blum and N. A. Blum, eds., *Sexual Selection and Reproductive Competition in Insects.* New York: Academic Press.

Bormann, F. H. 1961. Intraspecific root grafting and the survival of eastern white pine stumps. *Forest Sci.* 7:247-256.

Bormann, F. H. 1962. Root grafting and non-competitive relations between trees. Pp. 237-245 in T. T. Kozlowski, ed., *Tree Growth.* New York: Ronald Press.

Bormann, F. H. 1966. The structure, function and ecological significance of root grafts in *Pinus strobus* L. *Ecol. Monogr.* 36:1-26.

Bormann, F. H., and B. F. Graham. 1959. The occurrence of natural root grafting in eastern white pine, *Pinus strobus* L. and its ecological implications. *Ecology* 40:678-691.

Bosemark, N. O. 1956a. On accessory chromosomes in *Festuca pratensis.* III. Frequency and geographical distribution of plants with accessory chromosomes. *Hereditas* 42:189-210.

Bosemark, N. O. 1956b. Cytogenetics of accessory chromosomes in *Phleum* populations. *Hereditas* 42:443-466.

Bosemark, N. O. 1957. Further studies on accessory chromosomes in grasses. *Hereditas* 43:211-235.

Boucher, D. H. 1977. On wasting parental investment. *Amer. Natur.* 111:786-788.

Bower, F. O. 1882. The germination and embryogeny of *Gnetum gnemon. Quart. J. Micr. Sci.* 22:278-298.

Bowman, P., and A. McLaren. 1970. Cell number in early embryos from strains of mice selected for large and small body size. *Genet. Res.* 15:261-263.

Boyle, P., and J. Doyle. 1953. Development in *Podocarpus nivalis* in relation to other Podocarps. I. Gametophytes and fertilization. *Sci. Proc. Roy. Dublin Soc.* 26:179-205.

Bradley, M. V., and J. C. Crane. 1965. Supernumerary ovule development and parthenocarpy in *Ficus carica* L., var. King. *Phytomorphology* 15:85-92.

Bramlett, D. L. 1973. Pollen phenology and dispersal pattern for shortleaf pine in the Virginia piedmont. *USDA For. Serv. Res. Pap.* SE-104.

Bramlett, D. L., and T. W. Popham. 1971. Model relating unsound seed and embryonic lethal alleles in self-pollinated pines. *Silv. Genet.* 20:192-193.

Bray, O. E., J. J. Kennelly, and J. L. Guarino. 1975. Fertility of eggs produced on territories of vasectomized red-winged blackbirds. *Wilson Bull.* 87:187-195.

Brazeau, M., and J.-M. Veilleux. 1976. Bibliographie annotée sur les effets de la fertilisation sur la production de cones et de semences. *Québec, Ministère des Terres et Forêts, Service de la Recherche, Mémoire 25.*

Brennan, M., and J. Doyle. 1956. The gametophytes and embryogeny of *Arthrotaxis. Sci. Proc. Roy. Dublin Soc.* 27:193-252.

Brewbaker, J. L., and S. K. Majumder. 1961. Cultural studies of the pollen population effect and the self-incompatibility inhibition. *Amer. J. Bot.* 48:457-464.

Brewbaker, J. L., and B. H. Kwack. 1963. The essential role of calcium in pollen germination and pollen tube growth. *Amer. J. Bot.* 50:859-865.

Brink, R. A., and D. C. Cooper. 1940. Double fertilization and development of the seed in angiosperms. *Bot. Gaz.* 102:1-25.

Brink, R. A., and D. C. Cooper. 1947. The endosperm in seed development. *Bot. Rev.* 13:423-541.

Brough, M., and M. H. Taylor. 1940. An investigation of the life cycle of *Macrozamia spiralis* Miq. *Proc. Linn. Soc. N.S.W.* 65:494-524.

Brown, I. R. 1970. Seed production in Scots pine. Pp. 55-63 in L. C. Luckwill and C. V. Cutting, eds., *Physiology of Tree Crops*. London: Academic Press.

Brown, I. R. 1971. Flowering and seed production in grafted clones of Scots pine. *Silv. Genet.* 20:121-132.

Brownlie, G. 1953. Embryogeny of the New Zealand species of the genus *Podocarpus*, section Eupodocarpus. *Phytomorphology* 3:295-306.

Buchholz, J. T. 1918. Suspensor and early embryo of *Pinus*. *Bot. Gaz.* 66:185-228.

Buchholz, J. T. 1920a. Polyembryony among Abietineae. *Bot. Gaz.* 69:153-167.

Buchholz, J. T. 1920b. Embryo development and polyembryony in relation to the phylogeny of conifers. *Amer. J. Bot.* 7:125-145.

Buchholz, J. T. 1922. Developmental selection in vascular plants. *Bot. Gaz.* 73:249-286.

Buchholz, J. T. 1925. The embryogeny of *Cephalotaxus. fortunei*. *Bull. Torrey Bot. Club* 52:311-323.

Buchholz, J. T. 1926. Origin of cleavage polyembryony in conifers. *Bot. Gaz.* 81:55-71.

Buchholz, J. T. 1929. The embryogeny of the conifers. *Intern. Cong. Plant Sci.* 1:359-392.

Buchholz, J. T. 1930. The pine embryo and the embryos of related genera. *Trans. Ill. Acad. Sci.* 23:117-125.

Buchholz, J. T. 1931. The suspensor of *Sciadopitys*. *Bot. Gaz.* 92:243-262.

Buchholz, J. T. 1932a. The suspensor of *Cryptomeria japonica*. *Bot. Gaz.* 93:221-226.

Buchholz, J. T. 1932b. The embryogeny of *Chamaecyparis obtusa*. *Amer. J. Bot.* 19:230-238.

Buchholz, J. T. 1933. Determinate cleavage polyembryony, with special reference to *Dacrydium*. *Bot. Gaz.* 94:579-588.

Buchholz, J. T. 1936. Embryogeny of species of *Podocarpus* of the subgenus Stachycarpus. *Bot. Gaz.* 98:135-146.

Buchholz, J. T. 1939a. The morphology and embryogeny of *Sequoia gigantea*. *Amer. J. Bot.* 26:93-101.

Buchholz, J. T. 1939b. The embryogeny of *Sequoia sempervirens* with a comparison of the sequoias. *Amer. J. Bot.* 26:248-257.

Buchholz, J. T. 1940a. The embryogeny of *Torreya*, with a note on *Austrotaxus*. *Bull. Torrey Bot. Club* 67:731-754.

Buchholz, J. T. 1940b. The embryogeny of *Cunninghamia*. *Amer. J. Bot.* 27:877-883.

Buchholz, J. T. 1941a. Multi-seeded acorns. *Trans. Ill. Acad. Sci.* 34:99-101.

Buchholz, J. T. 1941b. Embryogeny of the Podocarpaceae. *Bot. Gaz.* 103:1-37.

Buchholz, J. T. 1942. A comparison of the embryogeny of *Picea* and *Abies*. *Madroño* 6:156-167.

Bull, J. J. 1980. Sex determination in reptiles. *Quart. Rev. Biol.* 55:3-21.

Bull, J. J. 1981. Sex ratio evolution when fitness varies. *Heredity* 46:9-26.

Bulmer, M. G., and P. D. Taylor. 1980. Dispersal and the sex ratio. *Nature* 284:448-449.

Burdon, R. D., and J. A. Zabkiewicz. 1973. Identical and non-identical seedling twins in *Pinus radiata*. *Can. J. Bot.* 51:2,001-2,004.

Burley, N. 1979. Clutch overlap and clutch size: alternative and complementary reproductive tactics. *Amer. Natur.* 115:223-246.

Burley, N. 1981a. Sex-ratio manipulation and selection for attractiveness. *Science* 211:721-722.

Burley, N. 1981b. Mate choice by multiple criteria in a monogamous species. *Amer. Natur.* 117:515-528.

Burley, N. 1982. Facultative sex-ratio manipulation. *Amer. Natur.* 120:81-107.

Burley, N. In press. The meaning of assortative mating. *Ethol. Sociobiol.*

Burlingame, L. L. 1913. The morphology of *Araucaria bra-*

siliensis. I. The staminate cone and female gametophyte. *Bot. Gaz.* 55:97-114.

Burlingame, L. L. 1914. The morphology of *Araucaria brasiliensis*. II. The ovulate cone and female gametophyte. *Bot. Gaz.* 57:490-508.

Burlingame, L. L. 1915. The morphology of *Araucaria brasiliensis*. III. Fertilization, the embryo, and the seed. *Bot. Gaz.* 59:1-39.

Buyak, A. V. 1975. Wood increment of *Picea abies* Karsten depending on intensity of seed-bearing. *Lesovedenie* (1975) (no. 5):58-62. (English summary.)

Cade, W. H. 1979. The evolution of alternative male reproductive strategies in field crickets. Pp. 343-379 in M. S. Blum and N. A. Blum, eds., *Sexual Selection and Reproductive Competition in the Insects*. New York: Academic Press.

Cade, W. H. 1981. Alternative male strategies: genetic differences in crickets. *Science* 212:563-564.

Cain, S. A., and L. G. Cain. 1944. Size-frequency studies of *Pinus palustris* pollen. *Ecology* 25:229-232.

Cain, S. A., and L. G. Cain. 1948. Size-frequency characteristics of *Pinus echinata*. *Bot. Gaz.* 110:325-330.

Caldwell, O. W. 1907. *Microcycas calocoma*. *Bot. Gaz.* 44:118-141.

Callan, E. McC. 1948. Effect of defoliation on reproduction of *Cordia macrostachya*. *Bull. Entom. Res.* 39:213-215.

Camefort, H. 1966a. Observation sur les mitochondries et les plastes d'origine pollinique après leur entrée dans une oosphère chez le Pin noir (*Pinus laricio* Poir. var. *austriaca* = *Pinus nigra* Arn.). *C. R. Acad. Sci. Paris* 263(D):959-962.

Camefort, H. 1966b. Etude en microscopie électronique de la dégénérescence du cytoplasme maternel dans les oosphères embryonnées du *Pinus laricio* Poir. var. *austriaca* (*P. nigra* Arn.). *C. R. Acad. Sci. Paris* 263(D):1,443-1,446.

Camefort, H. 1967. Fécondation et formation d'un néocytoplasme chez le *Larix decidua* Mill. (*Larix europea* D.C.). *C. R. Acad. Sci. Paris* 265(D):1,784-1,787.

Camefort, H. 1968. Sur l'organisation du néocytoplasme dans les proembryons tetranucléés du *Larix decidua* Mill. (*Larix europea* D.C.) et l'origine des mitochondries et des plastes de l'embryon chez cette espèce. *C. R. Acad. Sci. Paris* 266(D):88-91.

Camefort, H. 1969. Fécondation et proembryogénèse chez les Abiétacées (notion de néocytoplasme). *Rev. Cytol. Biol. Veg.* 32:253-271.

Canning, E. U., and K. Morgan. 1975. DNA synthesis, reduction, and elimination during life cycles of the Eimeriine Coccidian, *Eimeria tenella*, and the hemogregarine, *Hepatozoon domerguei*. *Exp. Parasitol.* 38:217-227.

Cantlon, J. E. 1970. The stability in natural populations and their sensitivity to technology. *Brookhaven Symp. Biol.* 22:197-203.

Capinera, J. L. 1979. Qualitative variation in plants and insects: effect of propagule size on ecological plasticity. *Amer. Natur.* 114:350-361.

Carbon, J. M. 1957. Shelterbelts and microclimate. *Forestry Commission Bull.* 29, Edinburgh.

Carl, C. M., and H. W. Yawney. 1972. Multiple seedlings in *Acer saccharum*. *Bull. Torrey Bot. Club* 99:142-144.

Castle, W. E. 1941. Size inheritance. *Amer. Natur.* 75:488-498.

Castle, W. E., and P. W. Gregory. 1929. The embryological basis of size inheritance in the rabbit. *J. Morph.* 48:81-104.

Cates, R. G. 1975. The interface between slugs and wild ginger: some evolutionary aspects. *Ecology* 56:391-400.

Cavalier-Smith, T. 1978. Nuclear volume control by nucleoskeletal DNA, selection for cell volume and cell growth rate, and the solution of the DNA C-value paradox. *J. Cell Sci.* 34:247-278.

Cech, F., J. Brown, and D. Weingartner. 1976. Wind damage to a yellow-poplar seed orchard. *Tree Planters' Notes* 27:3-4 (concl. on p. 20).

Chalupka, W. 1976. Effect of mineral fertilization on the content of reproductive organs in the litter dropped by a Norway spruce (*Picea abies* [L.] Karst.) canopy. *Arboretum Kórnickie* 21:333-359.

Chalupka, W., M. Giertych, and Z. Krolickowski. 1976. The effect of cone crops in Scots pine on tree diameter increment. *Arboretum Kórnickie* 21:361-366.

Chamberlain, C. J. 1910. Fertilization and embryogeny in *Dioon edule*. *Bot. Gaz.* 50:415-429.

Chamberlain, C. J. 1935. *Gymnosperms: Structure and Evolution.* Chicago: Univ. of Chicago Press (1957 Johnson reprint).

Champion, M. J., and G. S. Whitt. 1976. Synchronous allelic expression at the glucosephosphate isomerase *A* and *B* loci in interspecific sunfish hybrids. *Biochem. Genet.* 14:723-737.

Chandler, R. F. 1938. The influence of nitrogenous fertilizer application upon seed production of certain deciduous forest trees. *J. Forestry* 36:761-766.

Chaplin, S. J., and J. L. Walker. Ms. Floral display of a forest milkweed: energetic constraints and the adaptive value of the floral arrangement.

Charlesworth, B. 1978. Some models of the evolution of altruistic behaviour between siblings. *J. Theor. Biol.* 72:297-319.

Charlesworth, D., and B. Charlesworth. 1981. Allocation of resources to male and female functions in hermaphrodites. *Biol. J. Linn. Soc.* 15:57-74.

Charnov, E. L. 1976. Optimal foraging: attack strategy of a mantid. *Amer. Natur.* 110:141-151.

Charnov, E. L. 1979. Simultaneous hermaphroditism and sexual selection. *Proc. Nat. Acad. Sci.* (USA) 76:2,480-2,484.

Charnov, E. L., J. Maynard Smith, and J. J. Bull. 1976. Why be an hermaphrodite? *Nature* 263:125-126.

Charnov, E. L., R. L. Los-den Hartogh, W. T. Jones, and J. van den Assem. 1981. Sex ratio evolution in a variable environment. *Nature* 289:27-33.

Chatterjee, J., and H. Y. Mohan Ram. 1968. Gametophytes of *Equisetum ramosissimum* Desf. subsp. *ramosissimum*. I. Structure and development. *Botaniska Notiser* 121:471-490.

Chen, Z-K., and F-X. Wang. 1979. The gametophytes and fertilization in *Pseudotaxus*. *Acta Botanica Sinica* 21:19-29. (English summary.)

Chesnoy, L. 1969. Sur la participation du gamète mâle à la constitution du cytoplasme de l'embryon chez le *Biota orientalis* Endl. *Rev. Cytol. Biol. Veg.* 32:273-294.

Chesnoy, L. 1973. Sur l'origine paternelle des organites du proembryon du *Chamaecyparis lawsoniana* A. Murr. (Cupressacées). *Caryologia* 25 (Suppl.):223-232.

Chesnoy, L. 1977. Étude cytologique des gamètes, de la fécondation et de la proembryogenèse chez le *Biota orientalis* Endl. III. Fécondation et proembryogenèse; transmission du cytoplasme du gamète mâle au proembryon. *Rev. Cyt. Biol. Veget.* 40:293-396.

Chesnoy, L., and M. J. Thomas. 1971. Electron microscopy studies on gametogenesis and fertilization in gymnosperms. *Phytomorphology* 21:50-63.

Ching, K. K., and M. Simak. 1969. Competition among embryos in polyembryonal seeds of *Pinus sylvestris* and *Picea abies*. *XI Intern. Bot. Cong.* (abstr.):31.

Christiansen, H. 1969a. On the pollen grain and the fertilization mechanism of *Pseudotsuga menziesii* (Mirbel) Franco var. *viridis* Schwer. *Silv. Genet.* 18:97-104.

Christiansen, H. 1969b. On the germination of pollen of *Larix* and *Pseudotsuga* on artificial substrates, and on viability tests of pollen of coniferous forest trees. *Silv. Genet.* 18:104-107.

Christiansen, H. 1972a. On the development of pollen and the fertilization mechanism of *Picea abies* (L.) Karst. *Silv. Genet.* 21:51-61.

Christiansen, H. 1972b. On the development of pollen and the fertilization mechanisms of *Larix* and *Pseudotsuga menziesii. Silv. Genet.* 21:166-174.

Christiansen, H. 1973. On the anatomy of pollen grains of *Picea* and *Pinus. Silv. Genet.* 22:191-196.

Claassen, C. E. 1950. Natural and controlled crossing in safflower, *Carthamus tinctorius* L. *Agron. J.* 42:381-384.

Clare, T. S., and G. R. Johnstone. 1931. Polyembryony and germination of polyembryonic coniferous seeds. *Amer. J. Bot.* 18:674-683.

Clark, A. B. 1978. Sex ratio and local resource competition in a prosimian primate. *Science* 201:163-165.

Clausen, K. E. 1973. Genetics of yellow birch. *USDA For. Serv. Res. Pap.* WO-18.

Clay, K., and R. Shaw. 1981. An experimental demonstration of density-dependent reproduction in a natural population of *Diamorpha smallii,* a rare annual. *Oecologia* 51:1-6.

Cocucci, A. E. 1973. Some suggestions on the evolution of gametophytes of higher plants. *Phytomorphology* 23:109-124.

Coertze, A. F., H. P. van der Schijff, and H. G. Schweickerdt. 1971. Reproduction in *Podocarpus henkelii* Stapf ex Dallim. and Jacks. II. Embryogeny. *S. Afr. J. Sci.* 67:418-431.

Cohen, J. 1975. Gamete redundancy—wastage or selection? Pp. 99-112 in D. L. Mulcahy, ed., *Gamete Competition in Plants and Animals.* Amsterdam: North-Holland.

Cohen, J. 1977. *Reproduction.* London: Butterworths.

Coker, W. C. 1902. Notes on the gametophytes and embryo of *Podocarpus. Bot. Gaz.* 33:89-107.

Coker, W. C. 1903. On the gametophytes and embryo of *Taxodium. Bot. Gaz.* 36:1-27, 114-140.

Coker, W. C. 1907. Fertilization and embryogeny in *Cephalotaxus fortunei*. *Bot. Gaz.* 43:1-10.

Coles, J. F., and D. P. Fowler. 1976. Inbreeding in neighboring trees in two white spruce populations. *Silv. Genet.* 25:29-34.

Cook, M. T. 1903. Polyembryony in ginkgo. *Bot. Gaz.* 36:142.

Cook, P. L. 1939. A new type of embryogeny in the conifers. *Amer. J. Bot.* 26:138-143.

Cook, R. E. 1981. Plant parenthood. *Natural History* 90(7):30-35.

Cooke, F., P. J. Mirsky, and M. B. Seiger. 1972. Color preferences in the lesser snow goose and their possible role in mate selection. *Can. J. Zool.* 50:529-536.

Cooke, F., G. H. Finney, and R. F. Rockwell. 1976. Assortative mating in lesser snow geese (*Anser caerulescens*). *Behavior Genetics* 6:127-140.

Correns, C. 1921a. Versuche bei Pflanzen das Geschlechtsverhältnis zu verschieben. *Hereditas* 2:1-24.

Correns, C. 1921b. Zweite Fortsetzung der Versuche zur experimentellen Verschiebung des Geschlechtsverhältnisses. *Sitzungber. d. preussichen Akad. d. Wiss.*, pp. 330-354.

Correns, C. 1922. Geschlectsbestimmung und Zahlenverhältnis der Geschlechter beim Sauerrampfer (*Rumex acetosa*). *Biol. Zentralbl.* 42:465-480.

Correns, C. 1928. Bestimmung, Vererbung und Verteilung des Geschlechtes bei den höheren Pflanzen. *Handbuch der Vererbungwissenschaften* 2:1-138.

Cosmides, L. M., and J. Tooby. 1981. Cytoplasmic inheritance and intragenomic conflict. *J. Theor. Biol.* 89:83-129.

Coulter, J. M. 1908. The embryo sac and embryo of *Gnetum gnemon*. *Bot. Gaz.* 46:43-49.

Coulter, J. M., and W.J.G. Land. 1905. Gametophytes and embryo of *Torreya taxifolia*. *Bot. Gaz.* 39:161-178.

Craig, R. 1979. Parental manipulation, kin selection, and the evolution of altruism. *Evolution* 33:319-334.

Croker, T. C. 1964. Fruitfulness of longleaf trees more important than culture in cone yield. *J. Forestry* 62:822-823.

Crow, J. F. 1979. Genes that violate Mendel's rules. *Scientific American* 240(2):134-146.

Crowe, L. K. 1971. The polygenic control of outbreeding in *Borago officinalis. Heredity* 27:111-118.

Crozier, R. H. 1970. Coefficients of relationship and the identity of genes by descent in the Hymenoptera. *Amer. Natur.* 104:216-217.

Crozier, R. H. 1979. The genetics of sociality. Pp. 223-286 in H. R. Herman, ed., *Social Insects.* New York: Academic Press.

Crozier, R. H., and P. Pamilo. 1980. Asymmetry in relatedness: who is related to whom? *Nature* 283:604.

Cruden, R. W. 1977. Pollen:ovule ratios: a conservative indicator of breeding systems in flowering plants. *Evolution* 31:32-46.

Cruden, R. W., and S. M. Hermann-Parker. 1977. Temporal dioecism: an alternative to dioecism. *Evolution* 31:863-886.

Cutter, G. L., and E. T. Bingham. 1977. Effect of soybean male-sterile gene ms_1 on organization and function of the female gametophyte. *Crop Sci.* 17:760-764.

Dallimore, W., and A. B. Jackson. 1948. *A Handbook of Coniferae.* 3rd ed. London: Arnold.

D'Amato, F. 1977. *Nuclear Cytology in Relation to Development.* Cambridge: Cambridge Univ. Press.

Daniels, J. D. 1978. Efficacy of supplemental mass-pollination in a Douglas-fir seed orchard. *Silv. Genet.* 27:52-58.

Danill, D. M. 1932. *Macrocentrus ancylivorus* Rohwer, a polyembryonic braconid parasite of the oriental fruit moth.

N.Y. State Agr. Expt. Sta. Tech. Bull. 187, Geneva, New York.

Darwin, C. 1871. *The Descent of Man, and Selection in Relation to Sex.* London: John Murray.

Darwin, C. 1876. *The Effects of Cross and Self Fertilisation in the Vegetable Kingdom.* London: John Murray.

Daubenmire, R. 1960. A seven-year study of cone production as related to xylem layers and temperatures in *Pinus ponderosa. Am. Midl. Nat.* 64:187-193.

Davenport, R. 1979. *An Outline of Animal Development.* Reading, Mass.: Addison-Wesley.

Davidson, E. H. 1976. *Gene Activity in Early Development.* New York: Academic Press.

Davis, G. L. 1966. *Systematic Embryology of the Angiosperms.* New York: Wiley.

Dawkins, R. 1976. *The Selfish Gene.* New York: Oxford Univ. Press.

Dawkins, R. 1978. Replicator selection and the extended phenotype. *Z. Tierpsychol.* 47:61-76.

Dawkins, R., and T. R. Carlisle. 1976. Parental investment, mate desertion, and a fallacy. *Nature* 262:131-132.

Dawkins, R., and J. R. Krebs. 1979. Arms races between and within species. *Proc. Roy. Soc. Lond. B* 205:489-511.

Day, M. W., D. P. White, and J. W. Wright. 1972. Fertilizer application can improve red pine seed production. *Tree Planters' Notes* 23:25-27.

Dearborn, R. B. 1936. Nitrogen nutrition and chemical composition in relation to growth and fruiting of the cucumber plant. *Mem. Cornell Agric. Exp. Sta.* 192.

DeBarr, G. L., and L. R. Barber. 1975. Mortality factors reducing the 1967-1969 slash pine seed crop in Baker County, Florida — a life table approach. *USDA For. Serv. Res. Pap.* SE-131.

de Nettancourt, D. 1977. *Incompatibility in Angiosperms.* Berlin: Springer-Verlag.

Denison, N. P., and E. C. Franklin. 1975. Pollen management. *Bull. Brit. For. Comm.* 54:92-100.

deSilva, B.L.T., and M. S. Tambiah. 1952. A contribution to the life history of *Cycas rumphii* Miq. *Ceylon J. Sci.* (A) 12(5):223-244.

Detwiler, S. B. 1943. Better acorns from a heavily fertilized white oak tree. *J. Forestry* 41:915-916.

deWet, J.M.J., and H. T. Stalker. 1974. Gametophytic apomixis and evolution in plants. *Taxon* 23:689-697.

Dickmann, D. I., and T. T. Kozlowski. 1968. Mobilization by *Pinus resinosa* cones and shoots of C^{14}-photosynthate from needles of different ages. *Amer. J. Bot.* 55:900-906.

Dillman, A. C. 1938. Natural crossing in flax. *Agron. J.* 30:279-286.

Dixon, H. H. 1894. Fertilization in *Pinus sylvestris. Ann. Bot.* 8:21-34.

Doak, C. C. 1932. Multiple male cells in *Cupressus arizonica. Bot. Gaz.* 94:168-182.

Doak, C. C. 1937. Morphology of *Cupressus arizonica:* gametophytes and embryogeny. *Bot. Gaz.* 98:808-815.

Dodson, C. H. 1962. The importance of pollination in the evolution of the orchids of tropical America. *Amer. Orchid Soc. Bull.* 31:526-534, 641-649, 731-735.

Dogra, P. D. 1966a. Observations on *Abies pindrow* with a discussion on the question of occurrence of apomixis in gymnosperms. *Silv. Genet.* 15:11-20.

Dogra, P. D. 1966b. Embryogeny of the Taxodiaceae. *Phytomorphology* 16:125-141.

Doolittle, W. F., and C. Sapienza. 1980. Selfish genes, the phenotype paradigm and genome evolution. *Nature* 284:601-603.

Dorman, K. W. 1976. *The Genetics and Breeding of Southern Pines.* Agr. Handbook (USDA For. Serv.) 471.

Dorman, K. W., and A. E. Squillace. 1974. Genetics of slash pine. *USDA For. Serv. Res. Pap.* WO-20.

Dorman, K. W., and B. J. Zobel. 1973. Genetics of loblolly pine. *USDA For. Serv. Res. Pap.* WO-19.

Downie, D. G. 1928. Male gametophytes of *Microcycas calocoma. Bot. Gaz.* 85:437-450.

Downs, A. A., and W. E. McQuilkin. 1944. Seed production of southern Appalachian oaks. *J. Forestry* 42:913-920.

Doyle, J. 1954. Development in *Podocarpus nivalis* in relation to other podocarps. III. General conclusions. *Sci. Proc. Roy. Dublin Soc.* 26:347-377.

Doyle, J., and M. Brennan. 1971. Cleavage polyembryony in conifers and taxads—a survey. I. Podocarps, taxads, and taxodioids. *Sci. Proc. Roy. Dublin Soc.* A4:57-88.

Doyle, J., and M. Brennan. 1972. Cleavage polyembryony in conifers and taxads—a survey. II. Cupressaceae, Pinaceae, and conclusions. *Sci. Proc. Roy. Dublin Soc.* A4:137-158.

Doyle, J., and A. Kane. 1942. Pollination in *Tsuga pattoniana* and in species of *Abies* and *Picea. Sci. Proc. Roy. Dublin Soc.* 23:57-71.

Doyle, J., and M. O'Leary. 1935a. Pollination in *Saxegothaea. Sci. Proc. Roy. Dublin Soc.* 21:175-179.

Doyle, J., and M. O'Leary. 1935b. Pollination in *Tsuga, Cedrus, Pseudotsuga,* and *Larix. Sci. Proc. Roy. Dublin Soc.* 21:191-206.

Doyle, J. A. 1978. Origin of angiosperms. *Annu. Rev. Ecol. Syst.* 9:365-392.

Drayner, J. M. 1959. Self- and cross-fertility in field beans (*Vicia faba* Linn.). *J. Agric. Sci.* 53:387-403.

Duckett, J. G. 1975. Spermatogenesis in pteridophytes. Pp. 97-127 in J. G. Duckett and P. A. Racey, eds., *The Biology of the Male Gamete.* London: Academic Press.

Duckett, J. G., and P. R. Bell. 1971. Studies on fertilization in archegoniate plants. I. Changes in the structure of the spermatozoids of *Pteridium aquilinum* (L.) Kuhn during entry into the archegonium. *Cytobiologie* 4:421-436.

Duckett, J. G., and A. R. Duckett. 1980. Reproductive biology and population dynamics of wild gametophytes of *Equisetum*. *Bot. J. Linn. Soc.* 80:1-40.

D'Udine, B., and L. Partridge. 1981. Olfactory preferences of inbred mice (*Mus musculus*) for their own strain and for siblings: effects of strain, sex and cross-fostering. *Behaviour* 78:314-324.

Dunbar, A. 1975. Pollen development in *Eleocharis*. Pp. 57-58 in D. L. Mulcahy, ed., *Gamete Competition in Plants and Animals*. Amsterdam: North-Holland.

Dupler, A. W. 1917. The gametophytes of *Taxus canadensis* Marsh. *Bot. Gaz.* 64:115-164.

Durrant, A. 1962. The environmental induction of heritable change in *Linum*. *Heredity* 17:27-61.

Durrant, A., and T.W.A. Jones. 1971. Reversion of induced changes in amount of nuclear DNA in *Linum*. *Heredity* 27:431-439.

Eames, A. J. 1913. The morphology of *Agathis australis*. *Ann. Bot.* 27:1-38.

Eames, A. J. 1955. The seed and ginkgo. *J. Arnold Arb.* 36:165-170.

East, E. M. 1934. The nucleus-plasma problem. *Amer. Natur.* 68:289-303; 402-439.

Ebell, L. F. 1971. Girdling: its effect on carbohydrate status and on reproductive bud and cone development of Douglas fir. *Can. J. Bot.* 49:453-466.

Ebell, L. F. 1972. Cone-production and stem-growth response of Douglas fir to rate and frequency of nitrogen fertilization. *Can. J. For. Res.* 2:327-338.

Ebell, L. F., and E. E. McMullan. 1970. Nitrogenous substances associated with differential cone production responses of Douglas fir to ammonium and nitrate fertilization. *Can. J. Bot.* 48:2,169-2,177.

Eberhard, W. G. 1980. Evolutionary consequences of intracellular organelle competition. *Quart. Rev. Biol.* 55:231-249.

Edwards, G. A., and J. E. Endrizzi. 1975. Cell size, nuclear size, and DNA content relationships in *Gossypium. Can. J. Genet. Cytol.* 17:181-186.

Ehrendorfer, F. 1976. Evolutionary significance of chromosomal differentiation patterns in gymnosperms and primitive angiosperms. Pp. 220-240 in C. B. Beck, ed., *Origin and Early Evolution of Angiosperms.* New York: Columbia Univ. Press.

Ehrlich, P. R., and L. E. Gilbert. 1973. Population structure and dynamics of the tropical butterfly *Heliconius ethilla. Biotropica* 5:69-82.

Eickwort, K. R. 1973. Cannibalism and kin selection in *Labidomera clivicollis* (Coleoptera: Chrysomelidae). *Amer. Natur.* 107:452-453.

Eis, S., E. H. Garman, and L. F. Ebell. 1965. Relation between cone production and diameter increment of Douglas fir (*Pseudotsuga menziesii* [Mirb.] Franco), grand fir (*Abies grandis* [Dougl.] Lindl.), and western white pine (*Pinus monticola* [Dougl.]). *Can. J. Bot.* 43:1,553-1,559.

Eisenberg, J. 1966. The armadillo: a review of its natural history, ecology, anatomy, and reproductive physiology. *Handbuch der Zoologie* 10:1-92.

Elliott, C. G. 1948. The embryogeny of *Pherosphaera hookeriana. Proc. Linn. Soc. N.S.W.* 73:120-129.

Elliott, C. G. 1950. A further contribution on the life history of *Pherosphaera. Proc. Linn. Soc. N.S.W.* 75:320-333.

Endler, J. A. 1973. Gene flow and population differentiation. *Science* 179:243-250.

Endler, J. A. 1977. *Geographic Variation, Speciation, and Clines.* Monographs in Population Biology, No. 10. Princeton, N.J.: Princeton Univ. Press.

Endress, P. K. 1977. Evolutionary trends in the Hamamelidales-Fagales group. *Plant Syst. Evol. Suppl.* 1:321-347.

Engel, W. 1973. Onset of synthesis of mitochondrial enzymes during mouse development. *Humangenetik* 20:133-140.

Engel, W., and U. Wolf. 1971. Synchronous activation of the alleles coding for the S-form of the NADP-dependent malate dehydrogenase during mouse embryogenesis. *Humangenetik* 12:162-166.

Erdtman, G. 1954. *An Introduction to Pollen Analysis.* 2nd edition. Waltham, Mass.: Chronica Botanica.

Eriksson, G., B. Scholander, and V. Akebrand. 1973. Inbreeding depression in an old experimental plantation *Picea abies. Hereditas* 73:185-194.

Evans, G. M. 1968. Nuclear changes in flax. *Heredity* 23:25-38.

Evans, G. M., H. Rees, C. L. Snell, and S. Sun. 1972. The relationship between nuclear DNA amount and the duration of the mitotic cycle. *Chromosomes Today* 3:24-31.

Faegri, K., and J. Iverson. 1964. *Textbook of Pollen Analysis.* New York: Hafner.

Faegri, K., and L. van der Pijl. 1979. *The Principles of Pollination Ecology.* 3rd revised ed. Oxford: Pergamon Press.

Falconer, D. S. 1960. *Introduction to Quantitative Genetics.* Edinburgh: Oliver and Boyd.

Falconer, D. S. 1981. *Introduction to Quantitative Genetics.* 2nd ed. London: Longman.

Fankhauser, G. 1955. The role of nucleus and cytoplasm. Pp. 126-150 in B. H. Willier, P. A. Weiss, and V. Hamburger, eds., *Analysis of Development.* Philadelphia: Saunders.

Farmer, R. E., and G. C. Hall. 1971. Phenology and foliar control of floral morphogenesis in *Prunus serotina. Phytomorphology* 21:32-36.

Favre-Duchartre, M. 1956. Contribution à l'étude de la reproduction chez le *Ginkgo biloba. Rev. Cytol. Biol. Veg.* 17:1-218.

Favre-Duchartre, M. 1957. Contribution à l'étude de la reproduction chez *Cephalotaxus drupacea. Rev. Cytol. Biol. Veg.* 18:305-344.

Favre-Duchartre, M. 1958. Ginkgo, an oviparous plant. *Phytomorphology* 8:377-390.

Favre-Duchartre, M. 1962. Un mode de figuration des cycles biologiques végétaux appliqué à *Ginkgo, Araucaria, Taxus, Cephalotaxus,* et *Ephedra. Silv. Genet.* 11:16-19.

Favre-Duchartre, M. 1966. À propos de la fécondation, simple ou double, chez les plantes ovulées actuelles. *Ann. Sci. Natur. Botan.*, Paris, Ser. 12(7):421-444.

Favre-Duchartre, M. 1970. Des ovules aux graines. *Collection Monog. Botan. Biol. Veget.* 8 (Paris: Masson):1-136.

Favre-Duchartre, M. 1974. Phylogenetic aspects of the spermatophytes' double fertilization. Pp. 243-252 in H. F. Linskens, ed., *Fertilization in Higher Plants.* Amsterdam: North-Holland.

Fechner, G. H. 1972. Development of the pistillate flower of *Populus tremuloides* following controlled pollination. *Can. J. Bot.* 50:2,503-2,509.

Fechner, G. H. 1976. Development of unpollinated ovules of quaking aspen. *Proc. N.E. For. Tree Improv. Conf.* 23:121-128.

Ferguson, M. C. 1901. The development of the egg and fertilization in *Pinus strobus. Ann. Bot.* 15:435-479.

Ferguson, M. C. 1904. Contributions to the knowledge of the life history of *Pinus* with special reference to sporogenesis, the development of the gametophytes, and fertilization. *Proc. Wash. Acad. Sci.* 6:1-153.

Ferguson, M. C. 1913. Included cytoplasm in fertilization. *Bot. Gaz.* 56:501-502.

Fischer, E. A. 1981. Sexual allocation in a simultaneously hermaphroditic coral reef fish. *Amer. Natur.* 117:64-82.

Fisher, R. A. 1958. *The Genetical Theory of Natural Selection.* 2nd ed. New York: Dover.

Flesness, N. G. 1978. Kinship asymmetry in diploids. *Nature* 276:495-496.

Florence, R. G., and J. R. McWilliam. 1956. The influence of spacing on seed production. *Silv. Genet.* 5:97-102.

Foster, A. S., and E. M. Gifford. 1974. *Comparative Morphology of Vascular Plants.* San Francisco: Freeman.

Fowells, H. A., and G. H. Schubert. 1956. Seed crops of forest trees in the pine region of California. *USDA Tech. Bull.* 1,150.

Fowler, D. P. 1965a. Effects of inbreeding in red pine, *Pinus resinosa* Ait. II. Pollination studies. *Silv. Genet.* 14:12-23.

Fowler, D. P. 1965b. Effects of inbreeding in red pine, *Pinus resinosa* Ait. III. Factor affecting natural selfing. *Silv. Genet.* 14:37-46.

Fowler, D. P. 1965c. Effects of inbreeding in red pine, *Pinus resinosa* Ait. IV. Comparison with other northeastern *Pinus* species. *Silv. Genet.* 14:76-81.

Fowler, D. P., and D. T. Lester. 1970. The genetics of red pine. *USDA For. Serv. Res. Pap.* WO-8.

Frankie, G. W. 1975. Tropical forest phenology and pollinator plant coevolution. Pp. 192-209 in L. E. Gilbert and P. H. Raven, eds., *Coevolution of Animals and Plants.* Austin: Univ. of Texas Press.

Franklin, E. C. 1970. Survey of mutant forms and inbreeding depression in species of the family Pinaceae. *USDA For. Serv. Res. Pap.* SE-61.

Franklin, E. C. 1971. Estimates of frequency of natural selfing and of inbreeding coefficients in loblolly pine. *Silv. Genet.* 20:194-195.

Franklin, E. C. 1972. Genetic load in the loblolly pine. *Amer. Natur.* 106:262-265.

Free, J. B. 1970. *Insect Pollination of Crops.* New York: Academic Press.

Freeman, D. C., E. D. McArthur, K. T. Harper, and A. C. Blauer. 1981. Influence of environment on the floral sex ratio of monoecious plants. *Evolution* 35:194-197.

Frost, H. B. 1938. Nucellar embryony and juvenile characters in clonal varieties of citrus. *J. Heredity* 29:423-432.

Fröst, S. 1957. The inheritance of the accessory chromosomes in *Centaurea scabiosa. Hereditas* 43:405-421.

Fröst, S. 1959. The cytological behavior and mode of transmission of accessory chromosomes in *Plantago serraria. Hereditas* 45:191-210.

Funk, D. T. 1970. Genetics of black walnut. *USDA For. Serv. Res. Pap.* WO-10.

Furusato, K. 1953a. Studies on the polyembryony in *Citrus. Jap. J. Genet.* 28:165-166.

Furusato, K. 1953b. Studies on polyembryony in *Citrus. Nat. Inst. Genet. (Japan) Ann. Rep.* 5:56.

Furusato, K. 1960. Inheritance of mono- and polyembryony in *Citrus. Nat. Inst. Genet. (Japan) Ann. Rep.* 11:59-60.

Gabriel, W. J. 1967. Reproductive behavior in sugar maple: self-compatibility, cross-compatibility, agamospermy, and agamocarpy. *Silv. Genet.* 16:165-168.

Gadgil, M. 1972. Male dimorphism as a consequence of sexual selection. *Amer. Natur.* 106:574-580.

Gall, J. G. 1969. The genes for ribosomal RNA during oogenesis. *Genetics* (Suppl.) 61:121-132.

Ganders, F. R. 1979. The biology of heterostyly. *New Zealand J. Bot.*, pp. 607-635.

Gardner, G. 1977. The reproductive capacity of *Fraxinus excelsior* on the Derbyshire limestone. *J. Ecology* 65:107-118.

Gaümann, E. 1935. Der Stoffhaushalt der Buche (*Fagus sylvatica* L.) im Laufe eines Jahres. *Ber. Deut. Bot. Ges.* 63:366-377.

Geary, T. F. 1970. Direction and distance of pine pollen dispersal and seed orchard location on the copperbelt of Lambia. *Rhod. J. Agric. Res.* 8:123-130.

Gianordoli, M. 1964. Recherches cytologique sur la spermatogenèse, la fécondation et la proembryogenèse du

Sciadopitys verticillata. C. R. Acad. Sci. Paris 259:3,327-3,330.

Gianordoli, M. 1974. A cytological investigation on gametes and fecundation among *Cephalotaxus drupacea.* Pp. 221-232 in H. F. Linskens, ed., *Fertilization in Higher Plants.* Amsterdam: North-Holland.

Giddy, C. 1974. *Cycads of South Africa.* Cape Town: Purnell.

Giesel, J. T., and E. E. Zettler. 1980. Genetic correlations of life history parameters and certain fitness indices in *Drosophila melanogaster:* r_m, r_s, diet breadth. *Oecologia* 47:299-302.

Giesel, J. T., P. A. Murphy, and M. N. Manlove. 1982. The influence of temperature on genetic interrelationships of life history traits in a population of *Drosophila melanogaster:* what tangled data sets we weave. *Amer. Natur.* 119:464-479.

Gilissen, L.J.W.J., and H. F. Linskens. 1975. Pollen tube growth in styles of self-incompatible *Petunia* pollinated with radiated pollen and with foreign pollen mixtures. Pp. 201-205 in D. L. Mulcahy, ed., *Gamete Competition in Plants and Animals.* Amsterdam: North-Holland.

Gillham, N. W. 1969. Uniparental inheritance in *Chlamydomonas reinhardti. Amer. Natur.* 103:355-388.

Gillham, N. W. 1978. *Organelle Heredity.* New York: Raven Press.

Gleaves, J. T. 1973. Gene flow mediated by wind-borne pollen. *Heredity* 31:355-366.

Gloyne, R. W. 1954. Some effects of shelter belts upon local and microclimate. *J. Forestry* 27:85-95.

Godley, E. J. 1979. Floral biology in New Zealand. *New Zealand J. Bot.* 17:441-466.

Gould, S. J. 1977a. Caring groups and selfish genes. *Natural History* 86(10):20-24.

Gould, S. J. 1977b. *Ever Since Darwin.* New York: W. W. Norton.

Goyer, R. A., and L. H. Nachod. 1976. Loblolly pine conelet, cone, and seed losses to insects and other factors in a Louisiana seed orchard. *For. Sci.* 22:386-391.

Graham, B. F., and F. H. Bormann. 1966. Natural root grafts. *Bot. Rev.* 32:255-292.

Grano, C. X. 1957. Indices to potential cone production of loblolly pine. *J. Forestry* 55:890-891.

Grant, P. 1978. *Biology of Developing Systems.* New York: Holt, Rinehart, and Winston.

Grant, P. R. and T. D. Price. 1981. Population variation in continuously varying traits as an ecological genetics problem. *Amer. Zool.* 21:795-811.

Grant, V. 1975. *Genetics of Flowering Plants.* New York: Columbia Univ. Press.

Grant, V., and K. Grant. 1965. *Pollination in the Phlox Family.* New York: Columbia Univ. Press.

Gregory, P. H. 1973. *The Microbiology of the Atmosphere.* 2nd ed. New York: Wiley.

Gregory, P. W., and W. E. Castle. 1931. Further studies on the embryological basis of size inheritance in the rabbit. *J. Exp. Zool.* 59:199-211.

Grell, K. G. 1967. Sexual reproduction in protozoa. Pp. 147-213 in T.-T. Chen, ed., *Research in Protozoology.* Vol. 2. Oxford: Pergamon.

Grisez, T. J. 1975. Flowering and seed production in seven hardwood species. *USDA For. Serv. Res. Pap.* NE-315.

Gross, H. L. 1972. Crown deterioration and reduced growth associated with excessive seed production by birch. *Can. J. Bot.* 50:2,431-2,437.

Grun, P. 1976. *Cytoplasmic Genetics and Evolution.* New York: Columbia Univ. Press.

Grundwag, M. 1975. Seed set in some *Pistacia* L. (Anacardiaceae) species after inter- and intraspecific pollination. *Israel J. Bot.* 24:205-211.

Gunge, N., and Y. Nakatomi. 1972. Genetic mechanisms of rare matings of the yeast *Saccharomyces cerevisiae* heterozygous for mating type. *Genetics* 70:41-58.

Gurdon, J. B. 1974. *The Control of Gene Expression in Animal Development.* Oxford: Clarendon.

Gurdon, J. B., and H. R. Woodland. 1968. The cytoplasmic control of nuclear activity in animal development. *Biol. Rev.* 43:233-267.

Gutierrez, M. G., and G. F. Sprague. 1959. Randomness of mating in isolated polycross plantings of maize. *Genetics* 44:1,075-1,082.

Hagemann, R. 1976. Plastid distribution and plastid competition in higher plants and the induction of plastom mutations by nitrosourea compounds. Pp. 331-338 in T. Bucher, W. Neupert, W. Sebald, and S. Werner, eds., *Genetics and Biogenesis of Chloroplasts and Mitochondria.* Amsterdam: North-Holland.

Hagman, M. 1971. On self- and cross-incompatibility shown by *Betula verrucosa* Ehrh. and *Betula pubescens* Ehrh. *Commun. Inst. Forest. Fenn.* 73(6):1-125.

Hagman, M. 1975. Incompatibility in forest trees. *Proc. Roy. Soc. Lond. B* 188:313-326.

Hagman, M., and L. Mikkola. 1963. Observations on cross-, self-, and inter-specific pollinations in *Pinus peuce* Griseb. *Silv. Genet.* 12:73-79.

Haining, H. I. 1920. Development of embryo of *Gnetum. Bot. Gaz.* 70:436-445.

Hall, J. P., and I. R. Brown. 1976. Microsporogenesis, pollination, and potential yield of seed of *Larix* in NE Scotland. *Silv. Genet.* 25:132-137.

Hall, J. P. and I. R. Brown. 1977. Embryo development and yield of seed in *Larix. Silv. Genet.* 26:77-84.

Hamilton, W. D. 1963. The evolution of altruistic behavior. *Amer. Natur.* 97:354-356.

Hamilton, W. D. 1964a. The genetical evolution of social behaviour. I. *J. Theor. Biol.* 7:1-16.

Hamilton, W. D. 1964b. The genetical evolution of social behaviour. II. *J. Theor. Biol.* 7:17-52.

Hamilton, W. D. 1967. Extraordinary sex ratios. *Science* 156:477-488.

Hamilton, W. D. 1972. Altruism and related phenomena, mainly in social insects. *Annu. Rev. Ecol. Syst.* 3:193-232.

Hamilton, W. D. 1979. Wingless and fighting males in fig wasps and other insects. Pp. 167-220 in M. S. Blum and N. A. Blum, eds., *Sexual Selection and Reproductive Competition in Insects.* New York: Academic Press.

Hanover, J. W. 1975. Genetics of blue spruce. *USDA For. Serv. Res. Pap.* WO-28.

Harding, J., and C. L. Tucker. 1964. Quantitative studies on mating systems. I. Evidence for the non-randomness of outcrossing in *Phaseolus lunatus. Heredity* 19:369-381.

Harpending, H. C. 1979. The population genetics of interactions. *Amer. Natur.* 113:622-630.

Harper, J. H., and J. White. 1974. The demography of plants. *Annu. Rev. Ecol. Syst.* 5:420-463.

Hartley, C.W.S. 1970. Some environmental factors affecting flowering and fruiting in the oil palm. Pp. 269-285 in L. C. Luckwill and C. V. Cutting, eds., *Physiology of Tree Crops.* New York: Academic Press.

Hartnett, D. C., and W. G. Abrahamson. 1979. The effects of stem gall insects on life history patterns in *Solidago canadensis. Ecology* 60:910-917.

Hartung, J. 1977. An implication about human mating systems. *J. Theor. Biol.* 66:737-745.

Hartung, J. 1980. Parent-offspring conflict—a retraction. *J. Theor. Biol.* 87:815-816.

Hartung, J. 1981. Genome parliaments and sex with the red queen. Pp. 382-402 in R. D. Alexander and D. W. Tinkle, eds., *Natural Selection and Social Behavior: Recent Research and New Theory.* New York: Chiron Press.

Hashizume, H. 1973a. Fundamental studies on mating in

forest trees. V. Flowering and pollination in *Cryptomeria japonica*. *Bull. Fac. Agr. Tottori Univ.* 25:81-96.

Hashizume, H. 1973b. Fundamental studies on mating in forest trees. VI. Determination of favourable time for controlled pollination in *Cryptomeria japonica* and cone and seed yields in cross-, open-, and non-pollinations. *Bull. Fac. Agr. Tottori Univ.* 25:97-103. English summary.

Haupt, A. W. 1941. Oogenesis and fertilization in *Pinus lambertiana* and *P. monophylla*. *Bot. Gaz.* 102:482-498.

Hedegart, T. 1973. Pollination of teak (*Tectona grandis* L.). 2. *Silv. Genet.* 22:124-128.

Hedegart, T. 1976. Breeding systems, variation and genetic improvement of teak (*Tectona grandis* Lif.). Pp. 109-123 in J. Burley and B. T. Styles, eds., *Tropical Trees.* Linn. Soc. Symp., Ser. 2. New York: Academic Press.

Heinrich, B. 1975. Energetics of pollination. *Annu. Rev. Ecol. Syst.* 6:139-170.

Heinrich, B. 1979. Resource heterogeneity and patterns of movement in foraging bumblebees. *Oecologia* 14:235-245.

Herbers, J. M. 1979. The evolution of sex-ratio strategies in hymenopteran societies. *Amer. Natur.* 114:818-834.

Herbert, S. J. 1979. Density studies on lupins. I. Flower development. *Ann. Bot.* 43:55-63.

Heslop-Harrison, J. 1975. Male gametophyte selection and the pollen-stigma interaction. Pp. 177-199 in D. L. Mulcahy, ed., *Gamete Competition in Plants and Animals.* Amsterdam: North-Holland.

Hewitt, G. M., and F. M. Brown. 1970. The B-chromosome system of *Myrmeleotettix maculatus*. V. A steep cline in East Anglia. *Heredity* 25:363-371.

Hiirsalmi, H. 1969. *Trientalis europaea* L. A study of the reproductive biology, ecology, and variation in Finland. *Ann. Bot. Fenn.* 6:119-173.

Hill-Cottingham, D. G., and R. R. Williams. 1967. Effect

of time of application of fertilizer nitrogen on the growth, flower development, and fruit set of maiden apple trees, var. Lord Lambourne, and on the distribution of total nitrogen within the trees. *J. Hort. Sci.* 42:319-338.

Hinegardner, R. 1976. Evolution of genome size. Pp. 179-199 in F. J. Ayala, ed., *Molecular Evolution*. Sunderland, Mass: Sinauer.

Holloway, J. T. 1937. Ovule anatomy and development and embryogeny in *Phyllocladus alpinus* (Hook.) and in *P. glauca* (Carr.). *Trans. Proc. Roy. Soc. New Zealand* 67:149-165.

Horovitz, A., and J. Harding. 1972a. Genetics of *Lupinus*. 5: Intraspecific variability for reproductive traits in *Lupinus nanus*. *Bot. Gaz.* 133:155-165.

Horovitz, A., and J. Harding. 1972b. The concept of male outcrossing in hermaphrodite higher plants. *Heredity* 29:223-236.

Howe, H. F. 1976. Egg size, hatching asynchrony, sex, and brood reduction in the common grackle. *Ecology* 57:1,195-1,207.

Howell, D. J., and B. S. Roth. 1981. Sexual reproduction in agaves: the benefit of bats; the cost of semelparous advertising. *Ecology* 62:1-7.

Hutchinson, A. H. 1924. Embryogeny of *Abies*. *Bot. Gaz.* 77:280-289.

Istock, C. A. 1978. Fitness variation in a natural population. Pp. 171-190 in H. Dingle, ed., *Evolution of Insect Migration and Diapause*. New York: Springer-Verlag.

Jahromi, S. T., R. E. Goddard, and W. H. Smith. 1976. Genotype X fertilizer interactions in slash pine: growth and nutrient relations. *For. Sci.* 22:211-219.

Jain, S. K., and A. D. Bradshaw. 1966. Evolutionary divergence among adjacent plant populations. I. The evidence and its theoretical analysis. *Heredity* 21:407-441.

Janzen, D. H. 1971a. Seed predation by animals. *Annu. Rev. Ecol. Syst.* 2:465-492.

Janzen, D. H. 1971b. Euglossine bees as long-distance pollinators of tropical plants. *Science* 171:203-205.

Janzen, D. H. 1976. Effect of defoliation on fruit-bearing branches of the Kentucky coffee tree, *Gymnocladus dioicus* (Leguminosae). *Amer. Midl. Natur.* 95:474-478.

Janzen, D. H. 1977. A note on optimal mate selection by plants. *Amer. Natur.* 111:365-371.

Janzen, D. H., P. DeVries, D. E. Gladstone, M. L. Higgins, and T. M. Lewinsohn. 1980. Self- and cross-pollination of *Encyclia cordigera* (Orchidaceae) in Santa Rosa National Park, Costa Rica. *Biotropica* 12:72-74.

Jennings, D. L., and P. B. Topham. 1971. Some consequences of raspberry pollen dilution for its germination and for fruit development. *New Phytol.* 70:371-380.

Jensen, W. A. 1974. Reproduction in flowering plants. Pp. 481-503 in A. W. Robards, ed., *Dynamic Aspects of Plant Ultrastructure.* New York: McGraw-Hill.

Jerison, H. 1973. *Evolution of the Brain and Intelligence.* New York: Academic Press.

Johansen, D. A. 1950. *Plant Embryology.* Waltham, Mass.: Chronica Botanica.

John, B., and K. R. Lewis. 1975. *Chromosome Hierarchy.* London: Clarendon.

Johnson, C. M., D. L. Mulcahy, and W. C. Galinat. 1976. Male gametophyte in maize: Influences of the gametophytic genotype. *Theor. Appl. Genet.* 48:299-303.

Johnson, L. C. 1974. *Metasequoia glyptostroboides* Hu and Cheng. *USDA Agr. Handbook* 450:540-542.

Johnstone, G. R. 1940. Further studies on polyembryony and germination of polyembryonic pine seeds. *Amer. J. Bot.* 27:808-811.

Johri, B. M. 1963. Female gametophyte. Pp. 69-103 in P. Maheshwari, ed., *Recent Advances in the Embryology of*

Angiosperms. Internat. Soc. Plant Morphologists, Univ. of Delhi.

Jones, D. F. 1928. *Selective Fertilization.* Chicago: Univ. of Chicago Press.

Jones, M. D., and L. C. Newell. 1946. Pollination cycles and pollen dispersal in relation to grass improvement. *Nebraska Agric. Res. Bull.* 148.

Jones, M. D., and L. C. Newell. 1948. Size, variability, and identification of grass pollen. *J. Amer. Soc. Agron.* 40:136-143.

Jones, N. 1976. Some biological factors influencing seed setting in *Triplochiton scleroxylon* K. Schum. Pp. 125-134 in J. Burley and B. T. Styles, eds., *Tropical Trees.* Linn. Soc. Symp., Ser. 2. New York: Academic Press.

Jones, R. N. 1975. B-chromosome systems in flowering plants and animal species. *Int. Rev. Cytol.* 40:1-100.

Jones, R. N. 1976. Genome organization in higher plants. *Chromosomes Today* 5:117-130.

Jong, K. 1976. Cytology of the Dipterocarpaceae. Pp. 79-84 in J. Burley and B. T. Styles, eds., *Tropical Trees.* Linn. Soc. Symp., Ser. 2. New York: Academic Press.

Juliano, J. B. 1934. Origin of embryos in the strawberry mango. *Phil. J. Sci.* 54:553-559.

Juliano, J. B. 1937. Embryos of carabao mango (*Mangifera indica* Linn.). *Philip. Agricult.* 25:749-760.

Kaeiser, M. 1940. Morphology and embryogeny of the bald cypress. *Taxodium distichum* (L.) Rich. Ph.D. thesis, Univ. of Illinois.

Katsuta, M. 1971. Cone drop and pollination in *Pinus thunbergii* Parl. and *P. densiflora* Sieb. et Zucc. *Bull. Tokyo Univ. Forests* 65:87-106.

Kaur, A., C. O. Ha, K. Jong, V. E. Sands, H. T. Chan, E. Soepadmo, and P. S. Ashton. 1978. Apomixis may be widespread among trees of the climax rain forest. *Nature* 271:440-442.

Kawano, S., and S. Hayashi. 1977. Plasticity in growth and

reproductive energy allocation of *Coix Ma-yuen* Roman. cultivated at varying densities and nitrogen levels. *J. Coll. Liberal Arts, Toyama University* (Japan) 10:61-92.

Kayano, H. 1962. Cytogenetic studies in *Lilium callosum*. V. Supernumerary B chromosomes in wild populations. *Evolution* 16:246-253.

Keiding, H. 1968. Preliminary investigations of inbreeding and outcrossing in larch. *Silv. Genet.* 17:159-165.

Kellison, R. C., and B. J. Zobel. 1974. Genetics of Virginia pine. *USDA For. Serv. Res. Pap.* WO-21.

Khan, R. 1943. Contributions to the morphology of *Ephedra foliata* Boiss. II. Fertilization and embryogeny. *Proc. Nat. Acad. Sci. India*, Ser. B, 13:357-375.

Kildahl, N. J. 1908. The morphology of *Phyllocladus alpinus*. *Bot. Gaz.* 46:339-348.

Kirk, J.T.O., and R.A.E. Tilney-Bassett. 1978. *The Plastids.* 2nd ed. Amsterdam: Elsevier/North-Holland Biomedical Press.

Klekowski, E. J., Jr. 1972. Evidence against genetic self-incompatibility in homosporous fern *Pteridium aquilinum*. *Evolution* 26:66-73.

Klose, J., and U. Wolf. 1970. Transitional hemizygosity of the maternally derived allele at 6 PGD locus during early development of the cyprinid fish *Rutilus rutilus*. *Biochem. Genet.* 4:87-92.

Koller, D., and N. Roth. 1964. Studies in the ecological and physiological significance of amphicarpy in *Gymnarrhena micrantha* (Compositae). *Amer. J. Bot.* 51:26-35.

Konar, R. N. 1962. Some observations in the life history of *Pinus gerardiana* Wall. *Phytomorphology* 12:196-201.

Konar, R. N., and S. K. Banerjee. 1963. The morphology and embryology of *Cupressus funebris* Endl. *Phytomorphology* 13:321-328.

Konar, R. N., and Y. P. Oberoi. 1969a. Recent work on reproductive structures of living conifers and taxads—a review. *Bot. Rev.* 35:89-116.

Konar, R. N., and Y. P. Oberoi. 1969b. Studies on the morphology and embryology of *Podocarpus gracilior* Pilger. *Beitr. Biol. Pflanzen.* 45:329-376.

Konar, R. N., and S. Ramchandani. 1958. The morphology and embryology of *Pinus wallichiana* Jack. *Phytomorphology* 8:328-346.

Koski, V. 1971. Embryonic lethals of *Picea abies* and *Pinus sylvestris. Commun. Inst. Forest. Fenn.* 75(3):1-30.

Koski, V. 1973. On self-pollination, genetic load, and subsequent inbreeding in some conifers. *Commun. Inst. Forest. Fenn.* 78(10):1-42.

Kossuth, S. V., and G. H. Fechner. 1973. Incompatibility between *Picea pungens* Engelm. and *Picea engelmannii* Parry. *Forest Sci.* 19:50-60.

Kozlowski, T. T. 1971. *Growth and Development of Trees.* Vol. 2. New York: Academic Press.

Kozlowski, T. T., and J. C. Cooley. 1961. Root grafting in northern Wisconsin. *J. Forestry* 59:105-107.

Kozlowski, T. T., and T. Keller. 1966. Food relations of woody plants. *Bot. Rev.* 32:293-382.

Kramer, P. J., and T. T. Kozlowski. 1960. *Physiology of Trees.* New York: McGraw-Hill.

Kraus, J. F., and A. E. Squillace. 1964. Selfing vs. outcrossing under artificial conditions in *Pinus elliottii* Engelm. *Silv. Genet.* 13:72-76.

Kress, W. J. 1981. Sibling interactions and the evolution of pollen unit, ovule number, and pollen vector in angiosperms. *Syst. Bot.* 6:101-112.

Kriebel, H. B., and W. J. Gabriel. 1969. Genetics of sugar maple. *USDA For. Serv. Res. Pap.* WO-7.

Krugman, S. L., W. I. Stein, and D. M. Schmitt. 1974. Seed biology. USDA Agr. Handbook 450:5-40.

Kuijt, J. 1969. *The Biology of Parasitic Flowering Plants.* Berkeley: Univ. of California Press.

Labarca, C., and F. Loewus. 1973. The nutritional role of

pistil exudate in pollen tube wall formation in *Lilium longiflorum*. *Plant Physiol.* 52:87-92.

Labov, J. B. 1981. Pregnancy blocking in rodents: Adaptive advantages for females. *Amer. Natur.* 118:361-371.

Lahde, E., and K. Pahkala. 1974. Development and germination of the seeds of conifers according to literature. *Silva Fennica* 8:242-277. (English summary.)

Land, W.J.G. 1902. A morphological study of *Thuja*. *Bot. Gaz.* 34:249-259.

Land, W.J.G. 1907. Fertilization and embryogeny in *Ephedra trifurca*. *Bot. Gaz.* 44:273-292.

Langner, W. 1952. Eine Mendelspaltung bei *Auria*-formen von *Picea abies* (L.) Karst. als Mittel zur Klärung der Befruchtungsverhältnisse im Walde. *Z. Forstgenet.* 2:49-51.

Lanner, R. M. 1961. Living stumps in the Sierra Nevada. *Ecology* 42:170-173.

Lanner, R. M. 1966. Needed: a new approach to the study of pollen dispersion. *Silv. Genet.* 15:50-52.

Larson, M. M., and G. H. Schubert. 1970. Cone crops of ponderosa pine in central Arizona including the influence of Abert squirrels. *USDA For. Serv. Res. Pap.* RM-58.

Lawson, A. A. 1904a. The gametophytes, archegonia, fertilization, and embryo of *Sequoia sempervirens*. *Ann. Bot.* 18:1-28.

Lawson, A. A. 1904b. The gametophytes, fertilization, and embryo of *Cryptomeria japonica*. *Ann. Bot.* 18:417-444.

Lawson, A. A. 1907a. The gametophytes, fertilization, and embryo of *Cephalotaxus drupacea*. *Ann. Bot.* 21:1-23.

Lawson, A. A. 1907b. The gametophytes and embryo of the Cupressinae with special reference to *Libocedrus decurrens*. *Ann. Bot.* 21:281-301.

Lawson, A. A. 1909. The gametophytes and embryo of *Pseudotsuga douglasii*. *Ann. Bot.* 23:163-180.

Lawson, A. A. 1910. The gametophytes and embryo of *Sciadopitys verticillata*. *Ann. Bot.* 24:403-421.

Lawson, A. A. 1923a. The life history of *Microcachrys tetragona* (Hook.). *Proc. Linn. Soc. N.S.W.* 48:177-193.

Lawson, A. A. 1923b. The life-history of *Pherosphaera*. *Proc. Linn. Soc. N.S.W.* 48:499-516.

Lawson, A. A. 1926. A contribution to the life-history of *Bowenia*. *Trans. Roy. Soc. Edinburgh* 54:357-394.

Ledig, F. T., and J. H. Fryer. 1974. Genetics of pitch pine. *USDA For. Serv. Res. Pap.* WO-27.

Lee, C. K. 1955. Fertilization in *Ginkgo biloba*. *Bot. Gaz.* 117:79-100.

Lee, T. D. 1980. Extrinsic and intrinsic factors controlling reproduction in an annual plant. Ph.D. thesis, Univ. of Illinois.

Lee, T. D., and F. A. Bazzaz. 1980. Effects of defoliation and competition or growth and reproduction in the annual plant *Abutilon theophrasti*. *J. Ecology* 68:813-821.

Lee, T. D., and M. F. Willson. In press. Reproductive ecology of five herbs common in central Illinois. *Mich. Bot.*

Leigh, E. G., E. L. Charnov, and R. B. Warner. 1976. Sex ratio, sex change, and natural selection. *Proc. Nat. Acad. Sci.* (USA) 73:3,656-3,660.

Lemen, C. 1980. Allocation of reproductive effort to the male and female strategies in wind-pollinated plants. *Oecologia* 45:156-159.

Leonard, E. R. 1962. Inter-relations of vegetative and reproductive growth, with special reference to indeterminate plants. *Bot. Rev.* 28:353-410.

Le Tacon, F., and H. Oswald. 1977. Influence de la fertilisation minérale sur la fructification du hêtre (*Fagus silvatica*). *Ann. Sci. Forest.* 34:89-109.

Levin, D. A. 1975. Gametophytic selection in *Phlox*. Pp. 207-217 in D. L. Mulcahy, ed., *Gamete Competition in Plants and Animals*. Amsterdam: North-Holland.

Levin, D. A. 1979. Pollinator foraging behavior: genetic

implications for plants. Pp. 131-153 in O. T. Solbrig *et al.*, eds., *Topics in Plant Population Biology*. New York: Columbia Univ. Press.

Levin, D. A. 1981. Gene flow in seed plants revisited. *Ann. Missouri Bot. Gard.* 68:233-253.

Levin, D. A., and H. W. Kerster. 1974. Gene flow in seed plants. *Evol. Biol.* 7:139-220.

Lewin, B. 1974. *Gene Expression.* Vol. 2. London: Wiley.

Lewin, R. 1981. Do jumping genes make evolutionary leaps? *Science* 213:634-636.

Lewis, D. 1949. Incompatibility in flowering plants. *Biol. Rev.* 24:472-496.

Lewis, D. 1954. Comparative incompatibility in angiosperms and fungi. *Adv. Genet.* 6:235-285.

Lewis, D. 1979. Genetic versatility of incompatibility in plants. *New Zealand J. Bot.* 17:637-644.

Lewis, L. N., C. W. Coggins, Jr., and H. Z. Hield. 1964. The effect of biennial bearing and NAA on the carbohydrate and nitrogen composition of Wilking mandarin leaves. *Proc. Amer. Soc. Hort. Sci.* 84:147-157.

Li, C-C. 1955. *Population Genetics.* Chicago: Univ. of Chicago Press.

Lill, B. S. 1976. Ovule and seed development in *Pinus radiata*: postmeiotic development, fertilization, and embryogeny. *Can. J. Bot.* 54:2,141-2,154.

Lindgren, D. 1975. The relationship between self-fertilization, empty seeds and seeds originating from selfing as a consequence of polyembryony. *Studia Forest. Suecica* 126.

Linskens, H. F. 1969. Fertilization mechanisms in higher plants. Pp. 189-253 in C. B. Metz and A. Monroy, eds., *Fertilization.* Vol. 2. New York: Academic Press.

Lloyd, D. G. 1979. Some reproductive factors affecting the selection of self-fertilization in plants. *Amer. Natur.* 113:67-79.

Lloyd, D. G. 1980a. Sexual strategies in plants. I. An hypothesis of serial adjustment of maternal investment during one reproductive session. *New Phytol.* 86:69-79.

Lloyd, D. G. 1980b. Demographic factors and mating patterns in angiosperms. Pp. 67-88 in O. T. Solbrig, ed., *Demography and Evolution in Plant Populations.* Oxford: Blackwell's.

Lohani, D. N., and A. P. Kureel. 1973. Production of seed from tapped and untapped trees in chir pine (*Pinus roxburghii*, Sarg.). *Ind. Forester* 99:359-366.

Lombardo, G., and F. M. Gerola. 1968. Cytoplasmic inheritance and ultastructure of the male generative cell of higher plants. *Planta* 82:105-110.

Looby, W. J., and J. Doyle. 1937. Fertilization and proembryo formation in *Sequoia. Sci. Proc. Roy. Dublin Soc.* 21:457-476.

Looby, W. J., and J. Doyle. 1939. The ovule, gametophytes, and proembryo in *Saxegothea. Sci. Proc. Roy. Dublin Soc.* 22:95-117.

Looby, W. J., and J. Doyle. 1940. New observations in the life-history of *Callitris. Sci. Proc. Roy. Dublin Soc.* 22:241-255.

Looby, W. J., and J. Doyle. 1942. Formation of gynospore, female gametophyte, and archegonia in *Sequoia. Sci. Proc. Roy. Dublin Soc.* 23:35-54.

Looby, W. J., and J. Doyle. 1944a. The gametophyte of *Podocarpus andinus. Sci. Proc. Roy. Dublin Soc.* 23:222-237.

Looby, W. J., and J. Doyle. 1944b. Fertilization and early embryogeny in *Podocarpus andinus. Sci. Proc. Roy. Dublin Soc.* 23:257-270.

Lovett Doust, J., and J. L. Harper. 1980. The resource costs of gender and maternal support in an andromonoecious umbellifer, *Smyrnium olusatrum* L. *New Phytol.* 85:251-264.

Low, B. S. 1978. Environmental uncertainty and the parental strategies of marsupials and placentals. *Amer. Natur.* 112:197-213.

Luckwill, L. C. 1970. The control of growth and fruitfulness of apple trees. Pp. 237-253 in L. C. Luckwill and C. V. Cutting, eds., *Physiology of Tree Crops*. London: Academic Press.

Lundqvist, A. 1975. Complex self-incompatibility systems in angiosperms. *Proc. Roy. Soc. Lond. B* 188:235-245.

Lynch, M. Ms. Genomic incompatibility, general purpose genotypes, and geographic parthenogenesis.

Lyon, H. L. 1904. The embryogeny of ginkgo. *Minn. Bot. Stud.* 3:275-290.

Lyon, M. F. 1974. Review lecture: mechanisms and evolutionary origins of variable X-chromosome activity in mammals. *Proc. Roy. Soc. Lond. B* 187:243-268.

Lyons, L. A. 1956. The seed production capacity and efficiency of red pine cones (*Pinus resinosa* Ait.). *Can. J. Bot.* 34:27-36.

Maalkonen, E. 1971. Fertilizer treatment and seed crops of *Picea abies*. *Commun. Inst. Forest Fenn.* 73(4):1-16.

MacArthur, R. H. 1965. Ecological consequences of natural selection. Pp. 388-397 in T. H. Waterman and H. J. Morowitz, eds., *Theoretical and Mathematical Biology*. New York: Blaisdell.

MacArthur, R. H., and E. R. Pianka. 1966. On optimal use of a patchy environment. *Amer. Natur.* 100:603-609.

McCollum, J. P. 1934. Vegetative and reproductive responses associated with fruit development in the cucumber. *Mem. Cornell Agric. Exp. Station* 63.

MacKay, T.F.C. 1981. Genetic variation in varying environments. *Genet. Research, Camb.* 37:79-93.

McLemore, B. F. 1975. Cone and seed characteristics of fertilized and unfertilized longleaf pines. *USDA For. Serv. Res. Pap.* 50-109.

McLemore, B. F. 1977. Clone influences maturation of unpollinated strobili in southern pines. *Silv. Genet.* 26:134-135.

Macnair, M. R., and G. A. Parker. 1978. Models of parent-offspring conflict. II. Promiscuity. *Anim. Behav.* 26:111-122.

Macnair, M. R., and G. A. Parker. 1979. Models of parent-offspring conflict. III. Intra-brood conflict. *Anim. Behav.* 27:1,202-1,209.

McNeill, J., and C. W. Crompton. 1978. Pollen dimorphism in *Silene alba* (Caryophyllaceae). *Can. J. Bot.* 56:1,280-1,286.

McNeilly, T. 1968. Evolution in closely adjacent plant populations. II. *Agrostis tenuis* on a small copper mine. *Heredity* 23:99-108.

McWilliam, J. R., and F. Mergen. 1958. Cytology of fertilization in *Pinus. Bot. Gaz.* 119:246-249.

Magnuson, T., and C. J. Epstein. 1981. Genetic control of very early mammalian development. *Biol. Rev.* 56:369-408.

Maheshwari, P. 1950. *An Introduction to the Embryology of Angiosperms.* New York: McGraw-Hill.

Maheshwari, P., and R. N. Konar. 1971. *Pinus. Botan. Monogr.* (CSIR, India) 7:1-130.

Maheshwari, P., and R. C. Sachar. 1963. Polyembryony. Pp. 265-296 in P. Maheshwari, ed., *Recent Advances in the Embryology of Angiosperms.* Calcutta: Cambray.

Maheshwari, P., and M. Sanwal. 1963. The archegonium in gymnosperms: a review. *Memoirs Indian Bot. Soc.* 4:103-119.

Maheshwari, P., and H. Singh. 1967. The female gametophyte of gymnosperms. *Biol. Rev.* 42:88-130.

Malécot, G. 1948. *Les Mathématiques de l'Hérédité.* Paris: Masson.

Manes, C. 1975. Genetic and biochemical activities in preimplantation embryos. *Symp. Soc. Devel. Biol.* 33:133-163.

Markarian, D., and H. P. Olmo. 1959. Cytogenetics of *Rubus.* I. Reproductive behavior of *R. procera* Muell. *J. Heredity* 50:131-136.

Marshall, N. B. 1966. *The Life of Fishes.* Cleveland: World.

Martens, P. 1963. Recherches sur *Welwitschia mirabilis*—III. L'ovule et le sac embryonnaire. *La Cellule* 63:309-329.

Martin, P. C. 1950. A morphological comparison of *Biota* and *Thuja. Proc. Penn. Acad. Sci.* 24:65-112.

Mathews, A. C. 1939. The morphological and cytological development of the sporophylls and seed of *Juniperus virginiana* L. *J. Elisha Mitchell Sci. Soc.* 55:7-62.

Matthews, J. D. 1963. Factors affecting the production of seed by forest trees. *Forestry Abstr.* 24:i-xiii.

Matthews, J. D. 1970. Flowering and seed production in conifers. Pp. 45-53 in L. C. Luckwill and C. V. Cutting, eds., *Physiology of Tree Crops.* London: Academic Press.

Mattson, W. J. 1971. Relationship between cone crop size and cone damage by insects in red pine seed-production areas. *Can. Ent.* 103:617-621.

Mattson, W. J. 1978. The role of insects in the dynamics of cone production of red pine. *Oecologia* 33:327-349.

Maun, M. A., and P. B. Cavers. 1971. Seed production and dormancy in *Rumex crispus.* I. The effects of removal of cauline leaves at anthesis. *Can. J. Bot.* 49:1,123-1,130.

Maynard Smith, J. 1956. Fertility, mating behavior, and sexual selection in *Drosophila subobscura. J. Genetics* 54:261-279.

Maynard Smith, J. 1974. The theory of games and the evolution of animal conflict. *J. Theor. Biol.* 47:209-221.

Maynard Smith, J. 1978. *The Evolution of Sex.* Cambridge: Cambridge Univ. Press.

Maynard Smith, J., and G. R. Price. 1973. The logic of animal conflict. *Nature* 246:15-18.

Meagher, T. R., and J. Antonovics. 1982. Life history variation in dioecious plant populations: a case study of *Chamaelirium luteum.* In H. Dingle and J. Hegman, eds.,

Evolution and Genetics of Life Histories. New York: Springer-Verlag.

Mehra, P. N. 1950. Occurrence of hermaphrodite flowers and the development of female gametophyte in *Ephedra intermedia* Shrenk et Mey. *Ann. Bot. N.S.* 14:165-180.

Mehra, P. N., and R. K. Malhotra. 1947. Stages in the embryology of *Cupressus sempervirens* Linn. with particular reference to the occurrence of multiple male cells in the male gametophyte. *Proc. Nat. Acad. Sci. India,* Ser. B, 17:129-153.

Mehra, P. N., and M. K. Sircar. 1949. The structure and development of the female and male gametophytes in *Cupressus funebris. Proc. Nat. Inst. Sci. India* 15:15-23.

Mergen, F. 1976. Microsporogenesis and macrosporogenesis in *Pseudolarix amabilis. Silv. Genet.* 25:183-188.

Mergen, F., J. Burley, and G. M. Furnival. 1965. Embryo and seedling development in *Picea glauca* (Moench) Voss after self-, cross-, and wind-pollination. *Silv. Genet.* 14:188-194.

Metcalf, R. A., J. A. Stamps, and V. V. Krishnan. 1979. Parent-offspring conflict that is not limited by degree of kinship. *J. Theor. Biol.* 76:99-107.

Mikkelsen, V. M. 1949. Has temperature any influence on pollen size? *Physiol. Plant.* 2:323-324.

Mikkola, L. 1969. Observations on interspecific sterility in *Picea. Ann. Bot. Fenn.* 6:285-339.

Miyake, K. 1903. On the development of the sexual organs and fertilisation in *Picea excelsa. Ann. Bot.* 17:351-372.

Miyake, K. 1908. The development of gametophytes and embryogeny of *Cunninghamia. Bot. Mag. Tokyo* 22:45-50.

Miyake, K., and K. Yasui. 1911. On the gametophytes and embryo of *Pseudolarix. Ann. Bot.* 25:639-647.

Mochizuki, T. 1962. Studies on the elucidation of factors affecting the decline in tree vigor in apples as induced by fruit load. *Bull. For. Agr. Hirosaki Univ.* 8:40-124.

Mogensen, H. L. 1975. Ovule abortion in *Quercus* (Fagaceae). *Amer. J. Bot.* 62:160-165.

Moir, R. B., and D. P. Fox. 1977. Supernumerary chromosome distribution in provenances of *Picea sitchensis* (Bong.) Carr. *Silv. Genet.* 26:26-33.

Mooney, H. A., and R. I. Hays. 1973. Carbohydrate storage in two California Mediterranean-climate trees. *Flora* 162:295-304.

Morris, R. F. 1951. The effects of flowering on the foliage production and growth of balsam fir. *For. Chron.* 27:41-57.

Morton, N., J. Crow, and H. J. Muller. 1956. An estimate of the mutational damage in man from data on consanguinous marriges. *Proc. Nat. Acad. Sci.* (USA) 42:855-863.

Moseley, M. F. 1943. Contributions to the life history, morphology, and phylogeny of *Widdringtonia cupressoides*. *Lloydia* 6:109-132.

Mukai, T., H. Schaffer, and C. Cockerham. 1972. Genetic consequences of truncation selection at the phenotypic level in *Drosophila melanogaster*. *Genetics* 72:763-769.

Mulcahy, D. L. 1971. A correlation between gametophytic and sporophytic characteristics in *Zea mays* L. *Science* 171:1,155-1,156.

Mulcahy, D. L. 1974a. Correlation between speed of pollen tube growth and seedling height in *Zea mays* L. *Nature* 249:491-493.

Mulcahy, D. L. 1974b. Adaptive significance of gamete competition. Pp. 27-30 in H. F. Linskens, ed., *Fertilization in Higher Plants*. Amsterdam: North-Holland.

Mulcahy, D. L. 1979. The rise of the angiosperms: a genecological factor. *Science* 206:20-23.

Mulcahy, D. L., and G. B. Mulcahy. 1975. The influence of gametophytic competition in sporophytic quality in *Dianthus chinensis*. *Theor. Appl. Genet.* 46:277-280.

Mulcahy, D. L., G. B. Mulcahy, and E. Ottaviano. 1975.

Sporophytic expression of gametophytic competition in *Petunia hybrida*. Pp. 227-232 in D. L. Mulcahy, ed., *Gamete Competition in Plants and Animals*. Amsterdam: North-Holland.

Mulcahy, D. L., and S. M. Kaplan. 1979. Mendelian ratios despite nonrandom fertilization? *Amer. Natur.* 113:419-425.

Müntzing, A. 1928. Chromosome number, nuclear volume, and pollen grain size in *Galeopsis*. *Hereditas* 10:241-260.

Müntzing, A. 1954. Cytogenetics of accessory chromosomes (B-chromosomes). *Caryologia*, Suppl.:282-301.

Müntzing, A. 1957. Frequency of accessory chromosomes in rye strains from Iran and Korea. *Hereditas* 43:682-685.

Müntzing, A. 1966. Accessory chromosomes. *Bull. Bot. Soc. Bengal* 20:1-15.

Murneek, A. E. 1926. Effects of correlation between vegetative and reproductive functions in the tomato (*Lycopersicon esculentum* Mill.). *Plant Physiol.* 1:3-55.

Murneek, A. E. 1939. Some physiological factors in growth and reproduction of trees. *Amer. Soc. Hort. Sci. Proc.* 37:666-671.

Murrill, W. A. 1900. The development of the archegonium and fertilization in hemlock spruce (*Tsuga canadensis*, Carr.). *Ann. Bot.* 14:583-607.

Myers, J. H. 1981. Interaction between western tent caterpillars and wild rose: a test of some general plant herbivore hypotheses. *J. Anim. Ecol.* 50:11-25.

Myles, D. G. 1978. The fine structures of fertilization in the fern *Marsilea vestita*. *J. Cell Sci.* 30:265-281.

Nagl, W. 1974. Role of heterochromatin in the control of cell cycle duration. *Nature* 249:53-54.

Nagl, W., and F. Ehrendorfer. 1975. DNA content, heterochromatin, mitotic index, and growth in perennial and annual *Anthemidae* (Asteraceae). *Pl. Syst. Evol.* 123:35-54.

Nagl, W., B. Frisch, and E. Frolich. 1979. Extra-DNA during floral induction? *Pl. Syst. Evol.*, Suppl. 2:111-118.

Newman, H. H. 1914. Modes of inheritance in teleost hybrids. *J. Expt. Zool.* 16:447-490.

Nichols, J. D., and R. H. Chabreck. 1980. On the variability of alligator sex ratios. *Amer. Natur.* 116:125-137.

Nielsen, P. C., and M. Schaffalitzky de Muckadell. 1953. Flower observations and controlled pollinations in *Fagus*. *Silv. Genet.* 3:6-17.

Nienstaedt, H., and A. Teich. 1972. Genetics of white spruce. *USDA For. Serv. Res. Pap.* WO-15.

Nitsch, J. P. 1971. Perennation through seeds and other structures. Pp. 413-501 in F. S. Steward, ed., *Plant Physiology*. Vol. 6A. New York: Academic Press.

Nygren, A. 1967. Apomixis in the angiosperms. *Handbuch der Pflanzenphysiologie* 18:551-596.

Ockendon, D. J., and L. Currah. 1977. Self-pollen reduces the number of cross-pollen tubes in the styles of *Brassica oleracea* L. *New Phytol.* 78:675-680.

O'Connor, R. J. 1978. Brood reduction in birds: selection for fratricide, infanticide, and suicide? *Anim. Behav.* 26:79-96.

O'Gara, B. W. 1969. Unique aspects of the female pronghorn (*Antilocapra americana* Ord.). *Amer. J. Anat.* 125:217-232.

Ohba, K., M. Iwakawo, Y. Okada, and M. Murai. 1971. Paternal transmission of a plastid anomaly in some reciprocal crosses of sugi, *Cryptomeria japonica* D. Don. *Silv. Genet.* 20:101-107.

Ohno, S., C. Stenius, L. C. Christian, and C. Harris. 1968. Synchronous activation of both parental alleles at the 6-PGD locus of Japanese quail embryos. *Biochem. Genet.* 2:197-204.

Olive, P.J.W., and R. B. Clark. 1978. Physiology of reproduction. Pp. 271-368 in P. J. Mill, ed., *Physiology of Annelids*. New York: Academic Press.

Orgel, L. E., and F.H.C. Crick. 1980. Selfish DNA: the ultimate parasite. *Nature* 284:604-607.

Orians, G. H. 1961. The ecology of blackbird (*Agelaius*) social systems. *Ecol. Monogr.* 31:285-312.

Orians, G. H. 1980. Some Adaptations of Marsh-nesting Blackbirds. Monographs in Population Biology, No. 14. Princeton, N.J.: Princeton Univ. Press.

Orr-Ewing, A. L. 1957. A cytological study of the effects of self-pollination on *Pseudotsuga menziesii* (Mirb.) France. *Silv. Genet.* 6:179-185.

Orr-Ewing, A. L. 1976. Inbreeding Douglas-fir to the S_3 generation. *Silv. Genet.* 25:179-183.

Orr-Ewing, A. L. 1977. Female sterility in Douglas fir. *Silv. Genet.* 26:75-77.

Osborn, T.G.B. 1960. Some observations on the life-history of *Podocarpus falcatus*. *Aust. J. Bot.* 8:243-255.

Ottaviano, E., M. Sari-Gorla, and D. L. Mulcahy. 1975. Genetic and intergametophytic influences on pollen tube growth. Pp. 125-134 in D. L. Mulcahy, ed., *Gamete Competition in Plants and Animals.* Amsterdam: North-Holland.

Ottaviano, E., M. Sari-Gorla, and D. L. Mulcahy. 1980. Pollen tube growth rates in *Zea mays*: implications for genetic improvement of crops. *Science* 210:437-438.

Ottley, A. M. 1909. The development of the gametophytes and fertilization in *Juniperus communis* and *Juniperus virginiana. Bot. Gaz.* 48:31-46.

Owens, J. N., and M. Molder. 1973. A study of DNA and mitotic activity in the vegetative apex of Douglas fir during the annual growth cycle. *Can. J. Bot.* 51:1,395-1,409.

Owens, J. N., and M. Molder. 1975a. Pollination, female gametophyte, and embryo and seed development in yellow cedar (*Chamaecyparis nootkatensis*). *Can. J. Bot.* 53:186-199.

Owens, J. N., and M. Molder. 1975b. Sexual reproduction of mountain hemlock. *Can. J. Bot.* 53:1,811-1,826.

Owens, J. N., and M. Molder. 1977a. Seed-cone differentiation and sexual reproduction in western white pine (*Pinus monticola*). *Can. J. Bot.* 55:2,574-2,590.

Owens, J. N., and M. Molder. 1977b. Sexual reproduction of *Abies amabilis*. *Can. J. Bot.* 55:2,653-2,667.

Owens, J. N., and M. Molder. 1979a. Sexual reproduction of white spruce (*Picea glauca*). *Can. J. Bot.* 57:152-169.

Owens, J. N., and M. Molder. 1979b. Sexual reproduction of *Larix occidentalis*. *Can. J. Bot.* 57:2,673-2,690.

Owens, J. N., and M. Molder. 1980a. Sexual reproduction of sitka spruce (*Picea sitchensis*). *Can. J. Bot.* 58:886-901.

Owens, J. N., and M. Molder. 1980b. Sexual reproduction in western red cedar (*Thuja plicata*). *Can. J. Bot.* 58:1,376-1,393.

Pandey, K. K. 1979. Overcoming incompatibility and promoting genetic recombination in flowering plants. *New Zealand J. Bot.* 17:645-663.

Parker, G. A. 1974. Assessment strategy and the evolution of fighting behavior. *J. Theor. Biol.* 47:223-243.

Parker, G. A., R. R. Baker, and V.G.F. Smith. 1972. The origin and evolution of gamete dimorphism and the male-female phenomenon. *J. Theor. Biol.* 36:529-553.

Parker, G. A., and M. R. Macnair. 1978. Models of parent-offspring conflict. I. Monogamy. *Anim. Behav.* 26:97-110.

Parker, G. A., and M. R. Macnair. 1979. Models of parent-offspring conflict. IV. Suppression: evolutionary retaliation by the parent. *Anim. Behav.* 27:1,210-1,235.

Parsons, P. A. 1973. *Behavioural and Ecological Genetics: A study in "Drosophila."* Oxford: Clarendon Press.

Partridge, L. 1980. Mate choice increases a component of offspring fitness in fruit flies. *Nature* 283:290-291.

Patterson, J. T. 1927. Polyembryony in animals. *Quart. Rev. Biol.* 2:399-426.

Payne, R. B. 1979. Sexual selection and intersexual differences in variance of breeding success. *Amer. Natur.* 114:447-452.

Pearson, H.H.W. 1910. On the embryo of *Welwitschia*. *Ann. Bot.* 24:759-766.

Pederson, P. N., H. B. Johansen, and J. Jorgensen. 1961. Pollen spreading in diploid and tetraploid rye. II. Distance of pollen spreading and risk of intercrossing. Pp. 68-86 in *Royal Vet. Agric. Coll. Ann. Yearbook.*

Pessin, L. J. 1934. Effect of flower production on rate growth of vegetative shoots of longleaf pine. *Science* 80:363-364.

Pfahler, P. L. 1965. Fertilization ability of maize pollen grains. I. Pollen sources. *Genetics* 52:513-520.

Pfahler, P. L. 1967. Fertilization ability of maize pollen grains. II. Pollen genotype, female sporophyte and pollen storage interaction. *Genetics* 57:513-521.

Pfahler, P. L. 1975. Factors affecting male transmission in maize (*Zea mays*). Pp. 115-124 in D. L. Mulcahy, ed., *Gamete Competition in Plants and Animals.* Amsterdam: North-Holland.

Piesch, R. F., and R. F. Stettler. 1971. The detection of good selfers for haploid induction in Douglas-fir. *Silv. Genet.* 20:144-148.

Plancke, F. 1922. Samenerzeugung geharzter Fohren. *Forstwiss. Centr.* 44:172-175.

Platt, J. 1964. Strong inference. *Science* 146:347-353.

Player, G. 1979. Pollination and wind dispersal of pollen in *Arceuthobium*. *Ecol. Monogr.* 49:73-87.

Plym Forshell, C. 1974. Seed development after self-pollination and cross-pollination of Scots pine, *Pinus sylvestris* L. *Studia Forest. Suecica* 118:1-37.

Poddubmayer-Arnoldi, V. 1960. Study of fertilization in the living material of some angiosperms. *Phytomorphology* 10:185-198.

Policansky, D. 1981. Sex choice and the size advantage model in jack-in-the-pulpit (*Arisaema triphyllum*). *Proc. Nat. Acad. Sci.* (USA) 78:1,306-1,308.

Polis, G. A. 1980. The effect of cannibalism on the demography and activity of a natural population of desert scorpions. *Behav. Ecol. Sociobiol.* 7:25-35.

Polis, G. A. 1981. The evolution and dynamics of intraspecific predation. *Annu. Rev. Ecol. Syst.* 12:225-251.

Pomeroy, K. B., and C. F. Korstian. 1949. Further results on loblolly pine seed production and dispersal. *J. Forestry* 47:968-970.

Powell, G. R. 1977a. Patterns of development in *Abies balsamea* crowns and effects of megastrobilus production in shoots and buds. *Can. J. For. Res.* 7:498-509.

Powell, G. R. 1977b. Biennial strobilus production in balsam fir: a review of its morphogenesis and a discussion of its apparent physiological basis. *Can. J. For. Res.* 7:547-555.

Preer, J. R. 1971. Extrachromosomal inheritance: hereditary symbionts, mitochondria, chloroplasts. *Annu. Rev. Genet.* 5:361-406.

Price, H. J. 1976. Evolution of DNA content in higher plants. *Bot. Rev.* 42:27-52.

Price, H. J., and K. Bachmann. 1976. Mitotic cycle time and DNA in annual and perennial Microseridinae (Compositae, Cichoriaceae). *Plant. Syst. Evol.* 126:323-330.

Price, H. J., A. H. Sparrow, and A. F. Nauman. 1973. Correlations between nuclear volume, cell volume, and DNA content in meristematic cells of herbaceous angiosperms. *Experientia* 29:1,028-1,029.

Price, H. J., A. H. Sparrow, and A. F. Nauman. 1974. Evolutionary and developmental considerations of the variability of nuclear parameters in higher plants. I. Genome volume, interphase chromosome volume, and estimated DNA content of 236 gymnosperms. *Brookhaven Symp. Biol.* 25:390-421.

Price, M. V., and N. M. Waser. 1979. Pollen dispersal and optimal outcrossing in *Delphinium nelsoni*. *Nature* 277:294-297.

Priestley, C. A. 1970. Some observations on the effect of cropping on the carbohydrate content in trunks of apple trees over a long period. *Rep. E. Malling Res. Sta.* 1969:121-123.

Primack, R. B. 1978. Evolutionary aspects of wind pollination in the genus *Plantago* (Plantaginaceae). *New Phytol.* 81:449-458.

Primack, R. B., and D. G. Lloyd. 1980. Andromonoecy in the New Zealand montane shrub manuka, *Leptospermum scoparium* (Myrtaceae). *Amer. J. Bot.* 67:361-368.

Proctor, M., and P. Yeo. 1973. *The Pollination of Flowers.* London: Collins.

Puertas, M. J., and R. Carmona. 1976. Greater ability of pollen tube growth in rye plants with 2 B chromosomes. *Theor. Appl. Genet.* 47:41-43.

Pulliam, H. R. 1974. On the theory of optimal diets. *Amer. Natur.* 108:59-74.

Purseglove, J. W. 1972. *Tropical Crops. Monocotyledons.* Vol. 2. New York: Wiley.

Queller, D. In press. Kin selection and conflict in seed maturation. *J. Theor. Biol.*

Quinn, C. J. 1964. Gametophyte development and embryogeny in the Podocarpaceae. I. *Podocarpus,* Section Dacrycarpus. *Phytomorphology* 14:342-351.

Quinn, C. J. 1965. Gametophyte development and embryogeny in the Podocarpaceae. II. *Dacrydium laxifolium. Phytomorphology* 15:37-45.

Quinn, C. J. 1966a. Gametophyte development and embryogeny in the Podocarpaceae. III. *Dacrydium bidwel lii. Phytomorphology* 16:81-91.

Quinn, C. J. 1966b. Gametophyte development and embryogeny in the Podocarpaceae. IV. *Dacrydium colensoi:* general conclusions. *Phytomorphology* 16:199-211.

Rao, L. N. 1961. Life-history of *Cycas circinalis* L. *J. Indian Bot. Soc.* 40:601-619.

Raven, C. P. 1961. *Oogenesis: The Storage of Developmental Information.* New York: Pergaman.

Raynor, G. S., J. V. Hayes, and E. C. Odgen. 1970a. Experimental data on dispersion and deposition of timothy and corn pollen from known sources. *Brookhaven National Laboratory Bulletin* BNL 50266.

Raynor, G. S., E. C. Odgen, and J. V. Hayes. 1970b. Dispersion and deposition of ragweed pollen from experimental sources. *J. Appl. Meteorol.* 9:885-895.

Reed, F. C., and S. N. Stephenson. 1972. The effects of simulated herbivory on *Ambrosia artemisiifolia* L. and *Arctium minus* Schk. *Mich. Acad.* 4:359-364.

Reed, F. C., and S. N. Stephenson. 1973. Factors affecting seed number and size in burdock, *Arctium minus* Schk. *Mich. Acad.* 5:449-455.

Rees, H. 1972. DNA in higher plants. *Brookhaven Symp. Biol.* 23:394-417.

Reynolds, L. G. 1924. Female gametophyte of *Microcycas*. *Bot. Gaz.* 77:391-403.

Rick, C. M., M. Holle, and R. W. Thorp. 1978. Rates of cross-pollination in *Lycopersicon pimpinellifolium*: impact of genetic variation in floral characters. *Plant Syst. Evol.* 129:31-44.

Rickett, H. W. 1923. Fertilization in *Sphaerocarpus*. *Ann. Bot.* 37:225-256.

Robertson, A. 1904. Studies in the morphology of *Torreya californica*, Torrey. *New Phytol.* 3:205-216.

Rockwood, L. L. 1974. The effect of defoliation on seed production of six Costa Rican tree species. *Ecology* 54:1,363-1,369.

Rohlf, F. J., and G. D. Schnell. 1971. An investigation of the isolation by distance model. *Amer. Natur.* 108:649-664.

Rosen, W. G. 1975. Pollen pistil interactions. Pp. 153-164

in J. G. Duckett and P. A. Racey, eds., *Biology of the Male Gamete.* New York: Academic Press.

Roy Chowdhury, C. 1961. The morphology and embryology of *Cedrus deodara* (Roxb.) Loud. *Phytomorphology* 11:283-304.

Roy Chowdhury, C. 1962. The embryogeny of conifers: a review. *Phytomorphology* 12:313-338.

Russell, S. D. 1980. Participation of male cytoplasm during gamete fusion in an angiosperm, *Plumbago zeylandica. Science* 210:200-201.

Rutishauser, A. 1956. Cytogenetik des endosperms. *Ber. schweiz. bot. Gaz.* 66:318-336.

Rutishauser, A. 1960. Telocentric fragment chromosomes in *Trillium grandiflorum. Heredity* 15:241-246.

Sager, R. 1975. Patterns of inheritance of organelle genomes: molecular basis and evolutionary significance. Pp. 252-267 in C. W. Birky, P. S. Perlman, and T. J. Byers, eds., *Genetics and Biogenesis of Mitochondria and Chloroplasts.* Columbus: Ohio St. Univ. Press.

Sakai, K-I., H. Mukaide, and K. Tomita. 1968. Intraspecific competition in forest trees. *Silv. Genet.* 17:1-5.

Salisbury, E. J. 1942. *The Reproductive Capacity of Plants.* London: Bell.

Sanders, J.P.M., C. Heyting, and P. Borst. 1976. The variability of the mitochondrial genome of *Saccharomyces* strains. Pp. 511-517 in T. Bucher, W. Neupert, W. Sebald, and S. Werner, eds., *Genetics and Biogenesis of Chloroplasts and Mitochondria.* Amsterdam: North-Holland.

Sanwal, M. 1962. Morphology and embryology of *Gnetum gnemon* L. *Phytomorphology* 12:243-264.

Saran, S., and J.M.J. deWet. 1976. Environmental control of reproduction in *Dichanthium intermedium. J. Cytol. Genet.* 11:22-28.

Sari Gorla, M., E. Ottaviano, and D. Faini. 1975. Genetic

variability of gametophyte growth rate in maize. *Theor. Appl. Genet.* 46:289-294.

Sarvas, R. 1955. Investigations into the flowering and seed quality of forest trees. *Commun. Inst. For. Fenn.* 45(7):1-37.

Sarvas, R. 1962. Investigations on the flowering and seed crop of *Pinus silvestris. Comm. Inst. For. Fenn.* 53(4):1-198.

Sarvas, R. 1968. Investigations on the flowering and seed crop of *Picea abies. Comm. Inst. For. Fenn.* 67:1-84.

Saunier, R. E., and R. F. Wagle. 1965. Root grafting in *Quercus turbinella* Green. *Ecology* 46:749-750.

Saxton, W. T. 1909. Preliminary account of the ovule, gametophyte, and embryo of *Widdringtonia cupressoides. Bot. Gaz.* 48:161-178.

Saxton, W. T. 1910a. Contributions to the life history of *Widdringtonia cupressoides. Bot Gaz.* 50:31-48.

Saxton, W. T. 1910b. Contributions to the life-history of *Callitris. Ann. Bot.* 24:557-569.

Saxton, W. T. 1913a. Contributions to the life-history of *Actinostrobus pyramidalis,* Miq. *Ann. Bot.* 27:321-345.

Saxton, W. T. 1913b. Contributions to the life-history of *Tetraclinis articulata,* Masters, with some notes on the phylogeny of the Cupressoideae and Callitroidae. *Ann. Bot.* 27:577-606.

Saxton, W. T. 1934a. Notes on conifers. VIII. The Morphology of *Austrotaxus spicata* Compton. *Ann. Bot.* 48:411-427.

Saxton, W. T. 1934b. Notes on conifers. IX. The ovule and embryogeny of *Widdringtonia. Ann. Bot.* 48:429-431.

Saxton, W. T. 1936. Notes on conifers. X. Some normal and abnormal structures in *Taxus baccata. Ann. Bot.* 50:519-522.

Saxton, W. T., and J. Doyle. 1929. The ovule and gametophytes of *Arthrotaxus selaginoides,* Don. *Ann. Bot.* 43:833-840.

Schaal, B. A. 1980. Measurement of gene flow in *Lupinus texensis*. *Nature* 284:450-451.

Schaffer, W. M., and M. V. Schaffer. 1977. The adaptive significance of variation in reproductive habit in Agavaceae. Pp. 261-276 in B. Stonehouse and C. Perrins, eds., *Evolutionary Ecology*. London: Macmillan.

Schaffer, W. M., and M. V. Schaffer. 1979. The adaptive significance of variations in reproductive habit in the Agavaceae. II. Pollinator foraging behavior and selection for increased reproductive expenditure. *Ecology* 60:1,051-1,069.

Schemske, D. W. 1977. Flowering phenology and seed set in *Claytonia virginica* (Portulacaceae). *Bull. Torrey Bot. Club* 104:254-262.

Schemske, D. W. 1978. Evolution of reproductive characteristics in *Impatiens* (Balsaminaceae): the significance of cleistogamy and chasmogamy. *Ecology* 59:596-613.

Schemske, D. W. 1980a. Evolution of floral display in the orchid *Brassavola nodosa*. *Evolution* 34:489-493.

Schemske, D. W. 1980b. Floral ecology and hummingbird pollination of *Combretum farinosum* in Costa Rica. *Biotropica* 12:169-181.

Schemske, D. W., M. F. Willson, M. N. Melampy, L. J. Miller, L. Verner, K. M. Schemske, and L. B. Best. 1978. Flowering ecology of some spring woodland herbs. *Ecology* 59:351-366.

Schlising, R. A. 1976. Reproductive proficiency in *Paeonia californica* (Paeoniaceae). *Amer. J. Bot.* 63:1,095-1,103.

Schmitt, D. 1965. The pistillate inflorescence of sweetgum (*Liquidambar styraciflua* L.). *Silv. Genet.* 15:33-35.

Schoch-Bodmer, H. 1940. The influence of nutrition upon pollen grain-size in *Lythrum salicaria*. *J. Genet.* 40:393-402.

Schoen, D. J. 1977. Morphological, phenological, and pollen-distribution evidence of autogamy and xenogamy in *Gilia achilleifolia* (Polemoniaceae). *Syst. Bot.* 2:280-286.

Schoener, T. W. 1971. Theory of feeding strategies. *Annu. Rev. Ecol. Syst.* 2:369-404.

Schopf, J. M. 1943. The embryology of *Larix*. *Ill. Biol. Monogr.* 19(4):5-97.

Schultz, R. P. 1971. Stimulation of flower and seed production in a young slash pine orchard. *USDA For. Serv. Res. Pap.* SE-91.

Schwagmeyer, P. L. 1979. The Bruce effect: An evaluation of male/female advantages. *Amer. Natur.* 114:932-938.

Schwartz, D. 1971. Genetic control of alcohol dehydrogenase—a competition model for regulation of gene action. *Genetics* 67:411-425.

Searcy, W. A. 1979a. Female choice of mates: a general model for birds and its application to red-winged blackbirds (*Agelaius phoeniceus*). *Amer. Natur.* 114:77-100.

Searcy, W. A. 1979b. Male characteristics and pairing success in red-winged blackbirds. *Auk* 96:353-363.

Semler, D. E. 1971. Some aspects of adaptation in a polymorphism for breeding colours in the threespine stickleback (*Gasterosteus aculeatus*). *J. Zool. Lond.* 165:291-302.

Sharman, G. B. 1973. The chromosomes of non-eutherian mammals. Pp. 485-530 in A. B. Chiarelli and E. Capanna, eds., *Cytotaxonomy and Vertebrate Evolution*. London: Academic Press.

Shaw, W. R. 1896. Contribution to the life-history of *Sequoia sempervirens*. *Bot. Gaz.* 21:332-339.

Sheedy, G. 1974. Effets comparés de divers fertilisants sur production de semencer du sapin baumier. *Quebec Dept. of Lands and Forests, Res. Serv. Note* 3:1-11.

Sheppard, P. M. 1952. A note on non-random mating in the moth *Panaxia dominula* (L). *Heredity* 6:239-241.

Shoulders, E. 1967. Fertilizer application, inherent fruitfulness, and rainfall affect flowering of longleaf pine. *Forest Sci.* 13:376-383.

Shoulders, E. 1968. Fertilization increases longleaf and slash pine flower and cone crops in Louisiana. *J. Forestry* 66:193-197.

Shoulders, E. 1973. Rainfall influences female flowering of slash pine. *USDA For. Serv. Res. Note* SO-150.

Showalter, A. M. 1926. Studies in the cytology of the Anacrogynae. II. Fertilization in *Riccardia pinguis. Ann. Bot.* 40:713-726.

Showalter, A. M. 1927a. Studies in the cytology of the Anacrogynae. III. Fertilization in *Fossombronia angulosa. Ann. Bot.* 41:37-46.

Showalter, A. M. 1927b. Studies in the cytology of the Anacrogynae. IV. Fertilization in *Pellia fabbroniana. Ann. Bot.* 41:409-417.

Silander, J. A., and R. B. Primack. 1978. Pollination intensity and seed set in the evening primrose (*Oenothera fruticosa*). *Amer. Midl. Nat.* 100:213-216.

Silen, R. R. 1962. Pollen dispersal considerations for Douglas-fir. *J. Forestry* 60:790-795.

Silen, R. R. 1978. Genetics of Douglas-fir. *USDA For. Serv. Res. Pap.* WO-35.

Silvestri, F. 1937. Insect polyembryony and its general biological aspects. *Bull. Mus. Comp. Zool.* 81:469-498.

Simmonds, F. J. 1951. Further effects of the defoliation of *Cordia macrostachya* (Jacq.) R. and S. *Can. Entom.* 83:24-27.

Sinclair, T. R., and C. T. deWit. 1975. Photosynthate and nitrogen requirements for seed production by various crops. *Science* 189:565-567.

Singh, H. 1961. The life history and systematic position of *Cephalotaxus drupacea* Sieb. et Zucc. *Phytomorphology* 11:153-197.

Singh, H., and J. Chatterjee. 1963. A contribution to the

life history of *Cryptomeria japonica* D. Don. *Phytomorphology* 13:429-445.

Singh, H., and Y. P. Oberoi. 1962. A contribution to the life history of *Biota orientalis* Endl. *Phytomorphology* 12:373-393.

Singh, L. B. 1960. *The Mango.* New York: Interscience.

Sinnott, E. W. 1913. The morphology of the reproductive structures in the Podocarpinae. *Ann. Bot.* 27:39-82.

Smith, A. P. 1979. The paradox of autotoxicity in plants. *Evolutionary Theory* 4:173-180.

Smith, C. C. 1981. The facultative adjustment of sex ratio in lodgepole pine. *Amer. Natur.* 118:297-305.

Smith, R. W. 1923. Life history of *Cedrus atlantica. Bot. Gaz.* 75:203-208.

Smith, W. H., and R. G. Stanley. 1969. Cone distribution in crowns of slash pine (*Pinus elliottii* Engelm.) in relation to stem, crown, and wood increment. *Silv. Genet.* 18:86-91.

Snell, T. W., and D. G. Burch. 1975. The effects of density on resource partitioning in *Chamaesyce hirta* (Euphorbiaceae). *Ecology* 56:742-746.

Snyder, E. B. 1968. Seed yield and nursery performance of self-pollinated slash pine. *For. Sci.* 14:68-74.

Snyder, E. B., and A. E. Squillace. 1966. Cone and seed yields from controlled breeding of southern pines. *USDA For. Serv. Res. Pap.* SO-22.

Snyder, E. B., R. J. Dinus, and H. J. Derr. 1977. Genetics of longleaf pine. *USDA For. Res. Pap.* WO-33.

Socias i Company, R., D. E. Kester, and M. V. Bradley. 1976. Effects of temperature and genotype on pollen tube growth in some self-incompatible and self-compatible almond cultivars. *J. Amer. Soc. Hort. Sci.* 101:490-493.

Sokal, R., and F. Rohlf. 1969. *Biometry* San Francisco: W. H. Freeman.

Solbrig, O. T. 1976. On the relative advantages of cross- and self-fertilization. *Ann. Missouri Bot. Garden* 63:262-276.

Solbrig, O. T., and P. D. Cantino. 1975. Reproductive adaptations in *Prosopis* (Leguminosae, Mimosoideae). *J. Arnold Arb.* 56:185-210.

Sorensen, F. 1969. Embryonic genetic load in coastal Douglas-fir, *Pseudotsuga menziesii* var. *menziesii*. *Amer. Natur.* 103:389-398.

Sorensen, F. C. 1970. Self-fertility of a central Oregon source of ponderosa pine. *USDA For. Serv. Res. Pap.* PNW 109.

Sorensen, F. 1971. Estimate of self-fertility in coastal Douglas-fir from inbreeding studies. *Silv. Genet.* 20:115-120.

Sorensen, F. 1973. Frequency of seedlings from natural self-fertilization in coastal Douglas-fir. *Silv. Genet.* 22:20-24.

Sorensen, F. C. 1982. The roles of polyembryony and embryo viability in the genetic system of conifers. *Evolution* 36:725-733.

Sorensen, F. C., and R. S. Miles. 1974. Self-pollination effects on Douglas-fir and ponderosa pine seeds and seedlings. *Silv. Genet.* 23:135-138.

Sorensen, F. C., J. F. Franklin, and R. Woollard. 1976. Self-pollination effects on seed and seedling traits in noble fir. *For. Sci.* 22:155-159.

Sparnaaij, L. D., Y. O. Kho, and J. Baer. 1968. Investigation on seed production in tetraploid freesias. *Euphytica* 17:289-297.

Sparrow, A. H., V. Pond, and R. C. Sparrow. 1952. Distribution and behavior of supernumerary chromosomes during microsporogenesis in a population of *Trillium erectum* L. *Amer. Natur.* 86:277-292.

Sparrow, A. H., A. F. Rogers, and S. S. Schwemmer. 1968. Radio sensitivity studies with woody plants—I. Acute gamma irradiation survival data for 28 species and predictions for 190 species. *Radiation Botany* 8:149-186.

Squillace, A. E., and R. J. Bingham. 1958. Selective fertilization in *Pinus monticola* L. *Silv. Genet.* 7:188-196.

Stamps, J. A., R. A. Metcalf, and V. V. Krishnan. 1978. A genetic analysis of parent-offspring conflict. *Behav. Ecol. Sociobiol.* 3:369-392.

Stanlake, E. A., and J. N. Owens. 1974. Female gametophyte and embryo development in western hemlock (*Tsuga heterophylla*). *Can. J. Bot.* 52:885-893.

Stanley, R. G., and E. G. Kirby. 1973. Shedding of pollen and seeds. Pp. 295-340 in T. T. Kozlowski, ed., *Shedding of Plant Parts.* New York: Academic Press.

Starr, C. K. 1979. Origin and evolution of insect sociality: A review of modern theory. In H. R. Hermann, ed., *Social Insects.* Vol. I. New York: Academic Press.

Stebbins, G. L. 1950. *Variation and Evolution in Plants.* New York: Columbia Univ. Press.

Stebbins, G. L. 1976. Seeds, seedlings, and the origin of angiosperms. Pp. 300-311 in C. B. Beck, ed., *Origin and Early Evolution of Angiosperms.* New York: Columbia Univ. Press.

Steffan, K. 1963. Fertilization. Pp. 105-133 in P. Maheshwari, ed., *Recent Advances in the Embryology of Angiosperms.* Calcutta: Cambray.

Steinbrenner, E. C., J. W. Duffield, and R. K. Campbell. 1960. Increased cone production of young Douglas-fir following nitrogen and phosphorous fertilization. *J. Forestry* 58:105-109.

Stephens, S. G. 1956. The composition of an open pollinated segregating cotton population. *Amer. Natur.* 90:25-39.

Stephenson, A. G. 1979. An evolutionary examination of the floral display of *Catalpa speciosa* (Bignoniaceae). *Evolution* 33:1,200-1,209.

Stephenson, A. G. 1980. Fruit set, herbivory, fruit reduction, and the fruiting strategy of *Catalpa speciosa* (Bignoniaceae). *Ecology* 61:57-64.

Stephenson, A. G. 1981. Flower and fruit abortion: proximate causes and ultimate functions. *Annu. Rev. Ecol. Syst.* 12:253-279.

Sterling, C. 1948a. Gametophyte development in *Taxus cuspidata*. *Bull. Torrey Bot. Club* 75:147-165.

Sterling, C. 1948b. Abnormal prothallia in *Tsuga canadensis*. *Bot. Gaz.* 109:531-534.

Stern, K., and L. Roche. 1974. *Genetics of Forest Ecosystems*. London: Chapman and Hall.

Stiles, F. G. 1972. Time, energy, and territoriality in the Anna hummingbird (*Calypte anna*). *Science* 173:818-820.

Stiles, W. 1912. The Podocarpeae. *Ann. Bot.* 26:443-515.

Stockwell, W. P. 1939. Preembryonic selection in the pines. *J. Forestry* 37:541-543.

Strand, L. 1956. Pollen dispersal. *Silv. Genet.* 6:129-136.

Sugihara, Y. 1938. Fertilization and early embryogeny of *Chamaecyparis pisifera* S. Z. *Sci. Rep. Tohoku Univ.*, 4th ser., *Biol.* 13:9-14.

Sugihara, Y. 1939. Embryological observations on *Thujopsis dolabrata* var. Hondai Makino. *Sci. Rep. Tohoku Univ.*, 4th ser., *Biol.* 14:291-303.

Sugihara, Y. 1941. Embryological observations on *Taiwania cryptomeroides* Hayata. *Sci. Dep. Tokohu Imp. Univ. (Biol.)* 16:291-295.

Sugihara, Y. 1943a. The embryogeny of *Cunninghamia lanceolata* Hooker. *Sci. Rep. Tohoku Imp. Univ.* 16:187-192.

Sugihara, Y. 1943b. Embryological observations on *Keteleeria davidiana* Beissner var. *formosana* Hayata. *Sci. Rep. Tohoku Univ.*, 4th ser., *Biol.* 17:215-222.

Sugihara, Y. 1947a. The embryogeny of *Cryptomeria japonica* D. Don. *Bot. Mag. Tokyo* 60:47-52. (English summary.)

Sugihara, Y. 1947b. The embryogeny of *Cunninghamia kon-*

ishii Hayata. *Bot. Mag. Tokyo* 60:53-57. (English summary.)

Sugihara, Y. 1947c. The embryogeny of *Abies firma* Siebold et Zuccarini. *Bot. Mag. Tokyo* 60:58-62. (English summary.)

Sugihara, Y. 1956. The embryogeny of *Cupressus funebris* Endlicher. *Bot. Mag. Tokyo* 69:439-441.

Sugihara, Y. 1969. On the embryo of *Cryptomeria japonica. Phytomorphology* 19:110-111.

Sukhada, K., and Jayachandra. 1980. Pollen allelopathy—a new phenomenon. *New Phytol.* 84:739-746.

Sved, J. A., and F. J. Ayala. 1970. A population cage test for heterosis in *Drosophila pseudoobscura. Genetics* 66:97-113.

Swamy, B.G.L. 1973. Contributions to the monograph on *Gnetum.* I. Fertilization and proembryo. *Phytomorphology* 23:176-182.

Swamy, B.G.L., and K. V. Krishnamurthy. 1975. Embryo sac ontogenies in angiosperms—an elucidation. *Phytomorphology* 25:12-18.

Swanson, S. D., and S. H. Sohmer. 1976. The biology of *Podophyllum peltatum* L. (Berberidaceae), the may apple. II. The transfer of pollen and success of sexual reproduction. *Bull. Torrey Bot. Club* 103:223-226.

Sweet, G. B. 1973. Shedding of reproductive structures in forest trees. Pp. 341-382 in T. T. Kozlowski, ed., *Shedding of Plant Parts.* New York: Academic Press.

Tahara, M. 1940. Embryogeny of *Torreya nucifera* S. et Z. *Sci. Rep. Tohoku Imp. Univ.* 15:419-426.

Tamas, I. A., D. H. Wallace, P. M. Ludford, and J. L. Ozbun. 1979. Effect of older fruits on abortion and abscisic acid concentration of younger fruits in *Phaseolus vulgaris. Plant Physiol.* 64:620-672.

Tanaka, R. 1969. Deheterochromatinization of the chromosome in *Spiranthes sinensis. Japan. J. Genet.* 44:291-296.

Tang, S. H. 1948a. Observations on the embryogeny of *Juniperus chinensis. Bot. Bull. Acad. Sinica* 2:13-18.

Tang, S. H. 1948b. The embryogeny of *Torreya grandis. Bot. Bull. Acad. Sinica* 2:269-275.

Tanksley, S. D., D. Zamir, and C. M. Rick. 1981. Evidence for extensive overlap of sporophytic and gametophytic gene expression in *Lycopersicon esculentum. Science* 213:453-455.

Tappeiner, J. C. 1969. Effect of cone production on branch, needle, and xylem ring growth of Sierra Nevada Douglas-fir. *For. Sci.* 15:171-174.

Tauber, H. 1967. Differential pollen dispersion and filtration. Pp. 131-141 in E. J. Cushing and H. E. Wright, eds., *Quaternary Paleoecology*. New Haven, Conn.: Yale Univ. Press.

Taylor, P. D., and M. G. Bulmer. 1980. Local mate competition and the sex ratio. *J. Theor. Biol.* 86:409-419.

Ter-Avanesian, D. V. 1978. Significance of pollen amount for fertilization. *Bull. Torrey Bot. Club* 105:2-8.

Thiessen, D. D., and B. Gregg. 1980. Human assortative mating and genetic equilibrium: an evolutionary perspective. *Ethol. Sociobiol.* 1:111-140.

Thomas, M-J., and L. Chesnoy. 1969. Observations relatives aux mitochrondries Feulgen positive de la zone périnucléaire de l'oosphere du *Pseudotsuga menziesii* (Mirb.) Franco. *Rev. Cytol. Biol. Veg.* 32:165-182.

Thomson, J. D., and S.C.M. Barrett. 1981. Temporal variation of gender in *Aralia hispida* Vent. (Araliaceae). *Evolution* 35:1,094-1,107.

Thomson, J. D., and R. C. Plowright. 1980. Pollen carryover, nectar rewards, and pollinator behavior with special reference to *Diervilla lonicera. Oecologia* 46:68-74.

Threadgill, P. F., J. M. Baskin, and C. C. Baskin. 1981. The floral ecology of *Frasera caroliniensis* (Gentianaceae). *Bull. Torrey Bot. Club* 108:25-33.

Tilney-Bassett, R.A.E. 1975. Genetics of variegated plants. Pp. 268-308 in C. W. Birky, P. S. Perlam, and T. J. Byers, eds., *Genetics and Biogenesis of Mitochondria and Chloroplasts*. Columbus: Ohio St. Univ. Press.

Tilney-Bassett, R.A.E. 1976. The control of plastid inheritance in *Pelargonium*. IV. *Heredity* 37:95-107.

Tourte, Y. 1971. Quelques aspects cytologiques infrastructuraux au cours de la fécondation chez le *Pteridium aquilinum* L. *C. R. Acad. Sci. Paris* 272:1,236-1,239.

Traub, H. P. 1936. Artificial control of nucellar embryony in *Citrus*. *Science* 83:165-166.

Trivers, R. L. 1972. Parental investment and sexual selection. Pp. 136-179 in B. Campbell, ed., *Sexual Selection and the Descent of Man, 1871-1971*. Chicago: Aldine.

Trivers, R. L. 1974. Parent-offspring conflict. *Amer. Zool.* 14:249-264.

Trivers, R. L., and H. Hare. 1976. Haplodiploidy and the evolution of social insects. *Science* 191:249-263.

Trivers, R. L., and D. E. Willard. 1973. Natural selection of parental ability to vary the sex ratio of offspring. *Science* 179:90-91.

Troy, M. R., and D. E. Wimber. 1968. Evidence for a constancy of the DNA synthetic period between diploid-polyploid groups in plants. *Exp. Cell Res.* 53:145-154.

Udovic, D. 1981. Determinants of fruit set in *Yucca whipplei:* reproductive expenditure vs. pollinator availability. *Oecologia* 48:389-399.

Vaarama, A. 1953. Chromosome fragmentation and accessory chromosomes in *Orthotrichum tenellum*. *Hereditas* 39:305-316.

Valdeyron, G., B. Dommée, and P. Vernet. 1977. Self-fertilization in male-fertile plants of a gynodioecious species: *Thymus vulgaris* L. *Heredity* 39:243-249.

Van Haverbeke, D. F., and R. A. Read. 1976. Genetics of eastern red cedar. *USDA For. Serv. Res. Pap.* WO-32.

Van't Hof, J. 1965. Relationships between mitotic cycle duration, S period duration, and the average rate of DNA syntheses in the root meristem of several plants. *Exp. Cell Biol.* 39:48-58.

Van Winkle-Swift, K. P. 1977. Maturation of algal zygotes: alternative approaches for *Chlamydomonas reinhardtii* (Chlorophyceae). *J. Phycol.* 13:225-231.

Varnell, R. J. 1976. Cone and seed production in slash pine: effects of tree dimensions and climatic factors. *USDA For. Serv. Res. Pap.* SE-145.

Vasek, F. C., and J. Harding. 1976. Outcrossing in natural populations. V. *Evolution* 30:403-411.

Vasil, V. 1959. Morphology and embryology of *Gnetum ula* Brongn. *Phytomorphology* 9:167-215.

Vasil, V., and R. K. Sahni. 1964. Morphology and embryology of *Taxodium mucronatum* Tenore. *Phytomorphology* 14:369-384.

Vaughn, K. C., L. R. DeBonte, K. G. Wilson, and G. W. Schaeffer. 1980. Organelle alteration as a mechanism for maternal inheritance. *Science* 208:196-198.

Vazart, B. 1958. Différentiation de cellules sexuelles et fécondation chez les phanérogams. *Protoplasmatologia* 7(3a):1-158.

Verner, J. 1965. Selection for sex ratio. *Amer. Natur.* 99:419-421.

Vosa, C. G. 1972. Two track heredity: differentiation of male and female meiosis in *Thulbaghia. Caryologia* 25:275-281.

Vosa, C. G., and P. W. Barlow. 1972. Meiosis and B-chromosomes in *Listera ovata* (Orchidaceae). *Caryologia* 25:1-8.

Wade, M. J. 1979. Sexual selection and variance in reproductive success. *Amer. Natur.* 114:742-747.

Wade, M. J., and S. J. Arnold. 1980. The intensity of sexual selection in relation to male sexual behaviour, female choice, and sperm precedence. *Anim. Behav.* 28:446-461.

Wagenitz, G. 1955. Uber die Anderung der Pollengrösse von Getreiden durch verschiedene Ernährungsbedingungen. *Ber. Deutschen. Bot. Ges.* 68:297-302.

Wallace, B. 1958. The role of heterozygosity in *Drosophila* populations. *Proc. Tenth Internat. Cong. Genet.* 1:408-419.

Waller, D. M. 1980. Environmental determinants of outcrossing in *Impatiens capensis* (Balsaminaceae). *Evolution* 34:747-761.

Wang, C-W. 1970. Cone and seed production in controlled pollination of ponderosa pine. *Idaho Forest, Wildlife and Range Exp. Sta. Pap.* 7.

Wang, C-W. 1977. Genetics of ponderosa pine. *USDA For. Serv. Res. Pap.* WO-34.

Wang, C-W., T. O. Perry, and A. G. Johnson. 1960. Pollen dispersion of slash pine (*Pinus elliottii* Engelm.) with special reference to seed orchard management. *Silv. Genet.* 9:78-86.

Wang, F. H. 1948a. The early embryogeny of *Glyptostrobus*. *Bot. Bull. Acad. Sinica* 2:1-12.

Wang, F. H. 1948b. Life history of *Keteleeria*. I. Stobili, development of the gametophytes and fertilization in *Keteleeria evelyniana*. *Amer. J. Bot.* 35:21-27.

Wang, F. H. 1950. Observations on the embryogeny of *Podocarpus nagi*. *Bot. Bull. Acad. Sinica* 3:141-145.

Wang, F. H., and N. F. Chien. 1964. Embryogeny of *Metasequoia*. *Acta Bot. Sinica* 12:252-263.

Wardlaw, C. W. 1955. *Embryogenesis in Plants*. London: Methuen.

Wareing, P. F., and J. Patrick. 1975. Source-sink relations and the partition of assimilates in the plant. Pp. 481-499 in J. P. Cooper, ed., *Photosynthesis and Productivity in Different Environments*. Cambridge: Cambridge University Press.

Warner, R. R. 1980. The coevolution of behavioral and life history characteristics. Pp. 151-188 in G. W. Barlow and J. Silberberg, eds., *Sociobiology: Beyond Nature/Nurture?* Boulder, Colo.: Westview Press.

Waterkeyn, L. 1954. Études sur les Gnétales. I—Le strobile femelle, l'ovule et la graine. *La Cellule* 56:105-145.

Watkinson, A. R., and J. L. Harper. 1978. The demography of a sand dune annual: *Vulpia fascicalata* I. The natural regulation of populations. *J. Ecol.* 66:15-33.

Weatherhead, P. J., and R. J. Robertson. 1977. Harem size, territory quality, and reproductive success in the red-winged blackbird (*Agelaius phoeniceus*). *Can. J. Zool.* 55:1,261-1,267.

Weatherhead, P. J., and R. J. Robertson. 1979. Offspring quality and the polygyny threshold: "the sexy son hypothesis." *Amer. Natur.* 113:201-208.

Webber, J. M. 1940. Polyembryony. *Bot. Rev.* 6:575-598.

Wells, H. 1979. Self-fertilization: advantageous or deleterious? *Evolution* 33:252-255.

Wenger, K. F. 1953. The effect of fertilization and injury on the cone and seed production of loblolly pine seed trees. *J. Forestry* 51:570-573.

Wenger, K. F. 1954. The stimulation of loblolly pine seed trees by preharvest release. *J. Forestry* 52:115-118.

Wenger, K. F. 1957. Annual variation in the seed crops of loblolly pine. *J. Forestry* 55:567-569.

Werren, J. H. 1980. Sex ratio adaptations to local mate competition in a parasitic wasp. *Science* 208:1,157-1,159.

Werren, J. H., and E. L. Charnov. 1978. Facultative sex ratios and population dynamics. Nature 272:349-350.

West, J. D., W. J. Frels, V. M. Chapman. 1977. Preferential expression of the maternally derived X chromosome in the mouse yolk sac. *Cell* 12:873-882.

West Eberhard, M. J. 1975. The evolution of social behavior by kin selection. *Quart. Rev. Biol.* 50:1-33.

Westoby, M., and B. Rice. 1982. Evolution of the seed plants and inclusive fitness of plant tissues. *Evolution* 36:713-724.

White, M.J.D. 1973. *Animal Cytology and Evolution.* 3rd ed. Cambridge: Cambridge Univ. Press.

White, T. L., H. G. Harris, and R. C. Kellison. 1977. Conelet abortion in longleaf pine. *Can. J. For. Res.* 7:378-382.

Whitehead, D. R. 1969. Wind pollination in the angiosperms: evolutionary and environmental considerations. *Evolution* 23:28-35.

Whitehouse, H.L.K. 1959. Cross- and self-fertilization in plants. Pp. 207-261 in P. R. Bell, ed., *Darwin's Biological Work.* Cambridge: Cambridge Univ. Press.

Whitten, W. K., and C. P. Dagg. 1961. Influence of spermatozoa on the cleavage rate of mouse eggs. *J. Exp. Zool.* 148:173-183.

Wilcox, J. R., and K. A. Taft. 1969. Genetics of yellow-poplar (Liriodendron). *USDA For. Serv. Res. Pap.* WO-6.

Williams, G. C. 1966. *Adaptation and Natural Selection.* Princeton, N.J.: Princeton Univ. Press.

Williams, G. C. 1975. *Sex and Evolution.* Monographs in Population Biology, No. 8. Princeton, N.J.: Princeton Univ. Press.

Willson, M. F. 1972. Evolutionary ecology of plants: a review. I. Introduction and energy budgets. *Biologist* 54:140-147.

Willson, M. F. 1979. Sexual selection in plants. *Amer. Natur.* 113:777-790.

Willson, M. F. 1981. On the evolution of complex life cycles in plants: a review and ecological perspective. *Ann. Missouri Bot. Gard.* 68:275-300.

Willson, M. F., and P. W. Price. 1980. Resource limitation of fruit and seed production in some *Asclepias* species. *Can. J. Bot.* 58:2,229-2,233.

Willson, M. F., and D. W. Schemske. 1980. Pollinator limitation, fruit production, and floral display in pawpaw (*Asimina triloba*). *Bull. Torrey Bot. Club* 107:401-408.

Willson, M. F., and K. P. Ruppel. MS. Resource allocation and floral sex ratios in *Zizania palustris* L.

Willson, M. F., L. J. Miller, and B. J. Rathcke. 1979. Floral display in *Phlox* and *Geranium*: adaptive aspects. *Evolution* 33:52-63.

Wilson, E. B. 1925. *The Cell in Development and Heredity*. 3rd ed. New York: Macmillan.

Winton, L. L. 1968. Fertilization in forced quaking aspen and cottonwood. *Silv. Genet.* 17:20-21.

Wittenberger, J. F. 1976. The ecological factors selecting for polygyny in altricial birds. *Amer. Natur.* 110:779-799.

Wittwer, S. H. 1943. Growth hormone production during sexual reproduction of higher plants with special reference to synapsis and syngamy. *Res. Bull. Missouri Agric. Exp. Stat.* 371.

Wodehouse, P. R. 1935. *Pollen Grains: Their Structure, Identification, and Significance in Science and Medicine.* New York: McGraw-Hill.

Wold, M. L., and R. M. Lanner. 1965. New shoots from a 20-year-old swamp mahogany *Eucalyptus* stump. *Ecology* 46:755-756.

Wolf, L. L. 1969. Female territoriality in a tropical hummingbird. *Auk* 86:490-504.

Wolf, U., and W. Engel. 1972. Gene activation during early development of mammals. *Humangenetik* 15:99-118.

Wolfenbarger, D. O. 1946. Dispersion of small organisms, distance dispersion rates of bacteria, spores, seeds, pollen, and insects: incidence rates of diseases and injuries. *Amer. Midl. Nat.* 35:1-152.

Wolff, G. 1975. Investigations on pollen tube growth of *Pisum sativum* and some of its mutants. Pp. 161-175 in D. L. Mulcahy, ed., *Gamete Competition in Plants and Animals*. Amsterdam: North-Holland.

Wolgast, L. J., and B. B. Stout. 1977. Effects of age, stand density, and fertilizer application on bear oak reproduction. *J. Wildl. Manag.* 41:685-691.

Wright, J. W. 1952. Pollen dispersion of some forest trees. *Northeastern For. Exp. Sta. Pap.* 46.

Wright, S. 1922. Coefficients of inbreeding and relationship. *Amer. Natur.* 56:330-338.

Wright, S. 1943. Isolation by distance. *Genetics* 28:114-138.

Wright, S. 1946. Isolation by distance under diverse systems of mating. *Genetics* 31:39-59.

Wyatt, R. 1981. The reproductive biology of *Asclepias tuberosa*. II. Factors determining fruit-set. *New Phytol.* 88:375-385.

Wyatt, R., and R. L. Hellwig. 1979. Factors determining fruit set in heterostylous bluets, *Houstonia caerulea* (Rubiaceae). *Syst. Bot.* 4:103-114.

Yanai, U., and G. E. McClearn. 1972. Assortative mating in mice and the incest taboo. *Nature* 238:281-282.

Yanai, U., and G. E. McClearn. 1973. Assortative mating in mice. II. Strain differences in female mating preference, male preference, and the question of possible sexual selection. *Behav. Genet.* 3:65-74.

Yasukawa, K. 1981. Male quality and female choice of mate in the red-winged blackbird (*Agelaius phoeniceus*). *Ecology* 62:922-929.

Yocum, H. A. 1971. Releasing shortleaf pines increases cone and seed production. *USDA For. Serv. Res. Note* SO-125.

Young, M. S. 1910. The morphology of the Podocarpinae. *Bot. Gaz.* 50:81-100.

Yuasa, A. 1952. Studies in the cytology of the Pteridophyta. XXIX. Does the spermatozoid accompany the plastid in fertilization? *Cytologia* 16:347-351.

Zatykó, J. M., I. Simon, C. S. Szabó. 1975. Induction of polyembryony in cultivated ovules of red currant. *Plant Sci. Letters* 4:281-283.

Zeide, B. 1978. Reproductive behavior of plants in time. *Amer. Natur.* 112:636-639.

Zimmerman, M. 1980. Reproduction in *Polemonium*: competition for pollinators. *Ecology* 61:497-501.

Index

Genus/Species Index

Library of Congress Cataloging in Publication Data

Willson, Mary F.
Mate choice in plants.

(Monographs in population biology; 19)
Bibliography: p.
Includes index.
1. Plants, Sex in. I. Burley, Nancy. II. Title.
III. Series.
QK827.W646 1983 581.1'66 83-42590
ISBN 0-691-08333-9
ISBN 0-691-08334-7 (pbk.)